P9-DBL-596

About the Author

Paul Milo was born in 1969 and worked as a journalist and editor at various community newspapers for more than a decade, winning several awards during that time. Now working as a freelance writer, his work has appeared in *Editor and Publisher*, on Beliefnet, in *READ* magazine, and in other publications. He lives in Newark, New Jersey.

YOUR FLYING
CAR AWAITS

CONTENTS

In memory of my parents, Helen and Gabriel Milo

YOUR FLYING CAR AWAITS

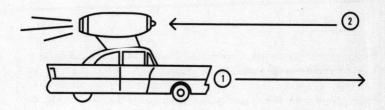

Robot Butlers, Lunar Vacations,
and Other Dead-Wrong Predictions
of the Twentieth Century

PAUL MILO

HARPER

NEW YORK • LONDON • TORONTO • SYDNEY

HARPER

YOUR FLYING CAR AWAITS. Copyright © 2009 by Paul Milo. All rights re-
served. Printed in the United States of America. No part of this book may
be used or reproduced in any manner whatsoever without written permis-
sion except in the case of brief quotations embodied in critical articles and
reviews. For information address HarperCollins Publishers, 10 East 53rd
Street, New York, NY 10022.

HarperCollins books may be purchased for educational, business, or
sales promotional use. For information please write: Special Markets Department, HarperCollins Publishers, 10 East 53rd Street, New York, NY
10022.

FIRST EDITION

Designed by Justin Dodd

Library of Congress Cataloging-in-Publication Data
  Milo, Paul.
    Your flying car awaits : robot butlers, lunar vacations, and other dead-
    wrong predictions from the twentieth century / Paul Milo. — 1st ed.
      p. cm.
    ISBN 978-0-06-172460-2
    1. Twentieth century—Forecasts. 2. Civilization, Modern—20th
  century—Miscellanea. I. Title.
  CB161.M483 2009
  909.82—dc22   2009019710

09 10 11 12 13   OV/RRD   10 9 8 7 6 5 4 3 2 1

INTRODUCTION

Whatever Happened to "the Future"?

A few years ago I came across an interesting article published in a London newspaper on New Year's Eve, 1899. In the article, the author assumes the perspective of a journalist writing exactly one hundred years later, looking back on the major scientific developments of the 20th century. Although the details escape me now, I remember being generally impressed with the writer's prescience. He wrote that in the previous ten decades the world's population grew to about 7 billion—not that far off from the actual number. He wrote that life expectancy in Britain and the other technologically advanced countries had risen to about seventy, again, close to being right on the money and all the more remarkable because back then the average person lived only into their mid-forties. He also described something like radio and television—not bad at a time when the most advanced form of communication was the telegraph.

But the writer also had a few goofs. For instance, he believed that here in the '00s, we'd still be making cross-country journeys in dirigible balloons. The British Empire would still be going strong. And there was no more war; he believed that something like a United Nations maintained something like a world police force that stepped in to stop armed conflicts before they started.

It was the author's misfires more than his spot-on guesses that struck a chord with me. When I was growing up in the 1970s, I can remember coming across the phrase "by the twenty-first century" quite a bit, as in, "by the twenty-first century, we'll all live to be one hundred and thirty" or "by the twenty-first century, we'll have computers that think." I found myself conducting a little mental experiment—if I were to travel back in time, say to my disco-era childhood, what would it be about the early 21st century that would most amaze people? What would it be about our own time that would seem the most "future-y" to people back then?

I'm sure they would be bowled over by, say, the cell phone, or the Internet. But for some reason, the more I thought about it, the more I came to feel that people in the 1970s would generally be underwhelmed with the world I came from. This is partly the fault of pop culture. Like a lot of '70s babies, one of my favorite TV shows was *The Six Million Dollar Man*, which was about a guy named Steve Austin who's endowed with superpowers after government doctors replace some of his body parts with robotic components. I suppose that as I was watching the show, a seed was planted in my young, still formative mind, because as I sit and think about it today, I can't help but feel more than a tinge of disappointment that we're nowhere near the future that the show envisioned.

Really, how come they can't give us bionic arms and legs, supervision and ultrasensitive hearing? There's been a lot of progress in surgery and a lot of progress in robotics since then, so why is it that I still can't bench-press a small car or jump a fifteen-foot fence like Austin could? The sci-fi of the '50s, '60s and '70s left millions of us expecting things to get a lot stranger over the next three decades than they actually did.

Still, I wasn't really interested in doing a book about the future as viewed through the lens of science fiction, primarily because of that second word: *fiction*. The creators of *The Six Million Dollar Man* set out to entertain, not predict. My latent disappointment in not having a bionic bod can't be blamed on them. But as I started to look into the subject a little more, I discovered that there was—and is—virtually an entire industry dedicated to telling us what the future held (or holds) according to people who claim to be intelligent enough to actually know. Think tanks, commentators, magazines, and professional associations, all purporting to give us an actual glimpse into the crystal ball. Once I started looking, I discovered a lot of material produced by various experts throughout the 20th century who made serious attempts to describe what the world would be like, or was supposed to be like, right now.

What struck me about many of these prognostications was that even though they used methods a lot more sophisticated than those employed by the creators of '70s-era television shows, many of them have proved to be almost as outlandish. By now, for instance, a fair number of experts believed that our main source of food would be plankton—thanks to what appeared to be looming food shortages. There were supposed to be moon bases today, too. And yes, of course, flying cars.

The more digging I did, the more I came across ideas and predictions that needed to be discussed. Many of these were ideas too amazing to leave between the dusty covers and professional journals of 20th century. The sheer wrongness of these guesses, I felt, made them too compelling to be forgotten. I wrote this book as a way to revisit those wacky, weird futures of the past.

The predictions in this book include many that were made by scientists and other experts with a cautious temperament, the kind of people who hedge their bets. For instance, the Hudson Institute, a think tank, produced an entire volume of forecasts in the mid-1960s that the authors, Herman Kahn and Anthony Wiener, carefully described as "projections," things that could happen but wouldn't necessarily happen. As I was doing my research, I imagined Kahn, Wiener, and some of the other futurists confronting me, protesting that their "forecasts" were intended only as a guide, only one of a range of "possible futures," as the writer Arthur C. Clarke, another one of my sources, once wrote in his own works of futurism.

While I can understand this argument, there are a few flaws with it. Many of the forecasts included here were the end result of extensive research by the forecasters making them. These weren't merely off-the-cuff guesses, and in many cases they were intended to be used as a reference by governments, universities, and corporations planning for the unforeseen future. These are not people making guesses with a toy from a Cracker Jack box; these are serious minds who were trying to use their expertise to decipher what the world would look like. In other words, I took these forecasts as seriously as they were intended to be taken. In addition, I'm pretty sure that the prognosticators we'll meet in this book would be more than happy to accept the credit for those forecasts that proved to be absolutely correct (and there's a section in this book listing some of their right-on-the-money guesses). If you aspire to guru status, you have to be as willing to accept a little mild ribbing for your mistakes as well as the accolades for your successes.

This book was a lot of fun to write. I loved the sheer zaniness of some of the predictions (like the guy in the 1940s who was sure we would all own personal helicopters by now), and the piquancy of others (like those who thought that in the 21st century we would no longer be fighting wars). I learned that predictions can say as much about the present as they do about the future (when times were good, as in the 1950s, optimism prevailed; in the '70s there was a lot more doom and gloom). I also learned that when trying to figure out what's coming next, it's generally better to err on the side of the gradual—especially when you're only looking ahead one generation (it takes time for truly sweeping changes to take root). I also learned that one of the worst mistakes any of us can make when trying to guess the future is to take a present-day trend and merely assume that it will continue.

Finally, I was reminded yet again that life is full of wild cards, that all kinds of things can happen that are simply unforeseeable. If nothing else, I hope readers of this book are reminded that trying to figure out what's around the bend ahead will always be an inexact science.

YOUR FLYING
CAR AWAITS

Chapter One

OUR BODIES, OURSELVES

In 1946, history's most famous clump of chicken cells finally died at the advanced age (for chicken cells, at least) of thirty-four.

The cells were kept alive at New York City's Rockefeller University by Alexis Carrel, a French-born surgeon and pioneer in organ transplantation, from 1912 until his death in 1944. Carrel, a highly respected medical researcher, was aiming to demonstrate that bodily tissues could live well beyond what was commonly perceived as their natural life span. For decades he assiduously drew the poisons from the cells and died believing that the same could one day be done for people. Researchers ended the experiment not because they had regarded it as a failure—after all, the cells had outlived the very person who had sustained them by two years—but because they were sure that the point had been made. Extending life, including human life, was a real possibility.

Two decades later other researchers discovered mistakes in Carrel's methods, realizing that every time he had drained the toxins from his geriatric tissues he was inadvertently replenishing the sample with fresh chicken cells. But by then the damage had already been done. While the cells were still alive, a newspaper, the *New York World Telegram*, annually reported the cells' "birthday" each January and other, sensationalized accounts claimed that Carrel was keeping a whole chicken heart beating. Carrel's work, flawed though it may have been, had helped inspire science to take over where Ponce de León had failed and so really find a fountain of youth (or at least a fountain of longevity).

And why not? Human life spans did rise dramatically across the world during the 20th century. An American female born in 1900, for example, lived to an average age of about fifty, but a typical girl born a century later can reasonably expect to be blowing out the candles on her eightieth birthday cake. The folks in the white lab coats were given much of the credit for this.

Doctors had indeed helped people live longer by 2000, by getting much better at treating injury and disease, encouraging sensible hygiene (countless millions were saved after late-19th century-surgeons simply began washing their hands before diving into a patient), and spreading the word about bad habits such as smoking. But the magical goal Carrel had sought still remains beyond science's grasp. We may live a lot longer than we used to, but nothing like the 150 or 200 years that some futurists believed we were going to achieve by now.

It's an appealing concept to think that science will allow us the advances to stave off life's most uncomfortable realities (death being the most uncomfortable reality of all), but of course the practical constraints of science are often out of sync with our visions for the future. The frailty of the human body makes it a particularly easy target when it comes to predictions. Every year we invest more and more money into the upkeep of our bodies. Every year we continue to wish and hope that science will make the impossible possible.

Even with all the medical advances that the 20th century saw, there were still predictions about the human body that far exceeded the realm of what was actually feasible. Along with endowing us with centuries of life, medical science was also expected to make those years much better, and in some cases a lot stranger, than they had ever been. Surveying the astounding advances in biology that had been made by then, writers in the 1950s and '60s looking ahead to the next forty years were absolutely giddy. There would be fantastic cures for previously untreatable diseases and methods for giving us green skin or fishlike gills.

Not everyone was so starry-eyed, however. When Watson and Crick discovered the double-helix shape of DNA in 1953, some researchers likened their achievement to opening Pandora's box: Once we understood the mechanics of human heredity, some feared it was a short leap to Huxley's *Brave New World*, a nightmare where genetically engineered elites would dominate society. Greater understanding of the brain could help in the fight against mental illness—and could potentially hand some 21st-century fascist government the perfect tool to keep the masses in line.

For centuries, each additional bit of knowledge we gained about our biology eventually led to an improvement in our quality of life. But as the 21st century drew closer, some feared that perhaps we had now learned too much for our own good. The predictions gathered here encompass the doom and gloom as well as the optimistic, providing a cross section of soothsayers who felt that for one reason or another medical science would change how our bodies function. When surveying the overblown fears and fondest hopes regarding what medicine today was to be like, however, it's probably best to begin at the beginning: childbirth.

Prediction: Baby Factories

In the 1920s the brilliant Scottish geneticist J. B. S. Haldane described a process he termed ectogenesis for creating human life, and it pretty much took all the fun out of making a baby. Instead of a man, a woman, and a bed (or a beach, or the backseat of a car), in the 21st century sperm would meet egg in a laboratory, and the developing embryo would grow to babyhood in a machine, or "artificial growth medium."

Haldane believed that by now only a small number of eccentric technophobes would still be "indulging" in traditional childbirth. Assuming the perspective of a fictional 21st-century essayist looking back on 20th-century history, Haldane wrote that France became the first country to adopt ectogenesis and that by 1968, "60,000 children annually [were produced] by this method." In England, "less than 30 percent of children are born of woman."

Haldane's work was picked up later in the century by several other researchers, including the Italian physician Daniele Petrucci, who in the 1960s fertilized a human egg outside the womb and kept the developing embryo alive long enough for arms, legs, and eyes to grow. In 1966, Petrucci assisted Soviet researchers in an unsuccessful attempt to bring a child to term using only artificial means. Petrucci's work, along with other advances, left many scientists and prognosticators convinced that the era of the laboratory baby would soon be upon us. Around the time of the Soviet experiment, the futurists Herman Kahn and Anthony Wiener described the possibility as "likely" to occur by 2000.

As if this *Matrix*-esque notion of millions of fetuses growing to maturity in a lab were not outlandish enough, there were also those who took this vision a step further, believing that those fetuses could also be customized according to the parents' wishes. A French scientist named Jean Rostand imagined a time when embryos would spend just the first few weeks in the uterus and finish their development in a device that he puckishly compared to a kangaroo's pouch. While the embryo grew, doctors could surgically alter its sex, eye color, or facial features. And since the baby no longer had to pass through the vaginal canal, its head could be much larger so as to accommodate an enhanced brain that had been suffused with additional neural cells. Care for a green-eyed boy supergenius? How about a brown-eyed girl with arms seven feet long?

Not surprisingly, there was a backlash. Petrucci was compared to Dr. Frankenstein and was rebuked by the Roman Catholic Church (an observant Catholic, Petrucci did cease his experiments for a while). Around the same time the magazine *New Scientist* proclaimed that we "are out of the realm of fancy now. Brave New World is on its way," in reference to Aldous Huxley's famous vision of Utopia gone wrong. Huxley's novel, which depicted a future of assembly-line baby factories and *mother* being a dirty word, is a satire. But by the 1960s, no one was laughing anymore.

So how close have we come to achieving Petrucci's and Haldane's goal? Well, that depends a bit on how you define *close*. Some might argue we are not far off. Test-tube babies have been around for more than thirty years. When Louise Brown was born in England to what had been an infertile couple in 1978, many believed that Haldane's prophecy, and one of the Catholic Church's worst fears, had been realized.

Meanwhile, baby incubators, machines that sustain premature infants until they're viable, have become much more advanced. Today it's not uncommon to save babies who are up to three months premature—and as of this writing, that threshold may have been pushed back even further. In February 2007, Amillia Taylor went home after being placed in an incubator at the age of just twenty-one weeks—the youngest surviving preemie to date. When she was born at a Florida hospital in October 2006, Amillia—like Louise Brown a test-tube baby—was less than ten inches long, about the size of a ballpoint pen. Conceived under glass and coming to term on a medical device, Amillia seems to come close to the "motherless" standard envisioned by Haldane eighty years ago.

Still, while the tools of test tubes and incubators have made striking advances, mankind has not put these instruments to the shocking uses that Haldane foresaw. There is nothing like a baby factory today, and as of this writing, nothing like it on the horizon either. Louise Brown, Amillia Taylor, and all their fellow in-vitros may have been conceived outside the womb, but once that process was complete, the tiny embryos were implanted back in a woman's body. Ectogenic babies, on the other hand, aren't born but are instead to be "decanted," plucked from a device where they've spent the entire development cycle. No human being has ever come into the world this way; everyone on the planet has spent a good spell inside a human uterus.

Though work on an artificial placenta is being carried out today, most researchers are focusing primarily on ways to assist infertile men and women, not to spare them from delivery altogether. This is partly because the dynamics of embryonic development are very complex; thus a machine that can do what an expectant mother's body does is probably a long way off (in 2003, writer Ronald Bailey described the artificial womb as one of those breakthroughs that are perpetually just a few years away). And no one really seems to be clamoring for a baby-making machine. Many women may want their labor to be as comfortable as science can make it, and technology may have helped millions of otherwise barren parents have kids, but few mothers appear to be asking science for an escape from the maternity ward.

Prediction: Giving Birth the Painless Way

The doctors who first tried to discover ways to alleviate labor pains two hundred years ago had to overcome many obstacles, including a big cultural one: the book of Genesis. The Bible tells us that when God expelled Adam and Eve from the Garden of Eden, He also attached other punishments to each of them suited to their particular crimes. And since it was Eve who committed the biggest no-no by eating the apple first and daring Adam to do the same, women got the shorter end of the stick. Not only would they forever be "subject" to their husbands, they were also doomed to suffer when giving birth: "Unto the woman He said, I will greatly multiply thy sorrow and thy conception; in sorrow thou shalt bring forth children. . . ."

This passage from Genesis was cited by commentators (almost all male, of course) who believed that God wanted giving birth to be an excruciating experience for the mom, and so the doctors who were trying to make the process a little less traumatic were therefore seen as interfering with the divine plan. Although it may sound downright bizarre—and painful!—to most modern ears, this debate was as divisive back then as today's arguments about creationism and evolution. Even now there are a handful of women who for religious reasons refuse any pain medication during childbirth. To them, the epidural is the devil's handiwork.

But the physicians who experimented with the first birth anesthetics had other, more concrete problems to contend with as well. Even as they tried chloroform, ether, and a cocktail of morphine and other drugs known as "Twilight Sleep" to ease the pain of birth, doctors worried about the effects these methods were having on the mother and child. And some believed, even well into the 20th century, that numbing any part of the woman's body could short-circuit the labor process altogether. These theories led more than one observer to believe that a practical birth anesthetic would never be found.

"In childbirth the woman under complete anesthesia has no labor and hence no child. Further, many anesthetic materials have a definite poisoning effect on the child, unused as it is to such things," wrote Yale professor C. C. Furnas in 1936, reporting what was then a common view. Eventually, of course, this prediction proved completely false as researchers came up with reliable methods to take some of the agony out of giving birth, while lingering taboos about what the Bible declared "natural" had largely melted away by the middle of the 20th century.

Prediction: Me, Me, Me—Human Cloning

While one of the main justifications for making babies in a lab was to spare women the discomfort of pregnancy, another purpose was to introduce a McDonald's-like efficiency to the haphazard process of passing on genes to the next generation, a way to crank out millions of guaranteed perfect babies. Cloning people, however, was meant to appeal to individuals whose idea of perfection could be found in the mirror.

Every human being who has ever lived gets his or her genes from a mother and a father. Clones are something else altogether. To see how, consider the most common way to create one: a process called nuclear transplantation. The nuclei of male sperm cells and female egg cells each contain chromosomes. When they come together under the right conditions, the two sets of chromosomes become one and the egg spontaneously begins to divide into more cells; an embryo has begun to develop. In nuclear transplantation, however, the nucleus of, say, the egg is removed and replaced with the nucleus of the sperm cell. When sperm from the male who contributed the transplanted nucleus now meets that egg, the embryo that's formed contains nothing but the genetic material of the male (the process can be adapted so females can be cloned, too). The resulting baby isn't the child of its father or mother but is more like his twin—a duplicate.

As far back as the 1950s, scientists had made a lot of progress in cloning frogs, sea urchins, and other relatively simple creatures. Scientists also tinkered with variations of this process, attempting to create offspring that shared, say, two-thirds of its mother's genes and one-third of its father's. One researcher, Landrum Shettles, even claimed in a 1979 interview that he knew precisely how to copy a human being.

The science behind cloning had advanced so rapidly that many respected researchers (including a few Nobel Prize winners) were sure that there would be lots of Multiple Me's running around by now. Around 1970, scientist Joshua Lederberg tentatively predicted the breakthrough to come within fifteen years. In the late 1960s, a Cambridge University physiologist imagined a time when a government agency would have been set up to issue "licenses" to individuals who wished to clone themselves. Writer Gordon Rattray Taylor, commenting on this possibility, observed that "23 Mozarts would scarcely be tolerable: 23 Hitlers or Stalins at once hardly bear thinking about."

Rattray Taylor had named one of the greatest fears surrounding cloning: that it would gradually lead to eugenics, in which only those deemed fit would be permitted to literally repeat themselves. Others worried about cloned people becoming "commodified," bred to provide their "owners" with replacement organs (an idea explored in the 2005 film *The Island*). Still others, meanwhile, foresaw cloning and ectogenesis together destroying the mother-child bond. "Essentially motherhood will be abolished. . . . The word 'mother' may or may not persist, but its essence . . . will be gone," one writer predicted in 1969.

The possibility of human cloning shows that there is sometimes a vast gulf between what we could do and what we would do. When Dolly the sheep was born in 1996, even nonscientists immediately grasped the implications and were appalled: cloning salamanders and sea urchins was one thing, but making carbon copies of mammals that are very similar to us genetically was something else. Calls came from many quarters—including from the very scientists who had helped make Dolly—for an outright ban on human cloning. In 2005, the United Nations did pass a nonbinding resolution asking its member countries to outlaw any attempts to create a baby that was a duplicate of another human being. (Some governments would allow cloning to create embryos as a source of stem cells—material that scientists believe could be used to create a supply of tissues for donation—but those embryos would be destroyed long before they develop beyond a very simple stage.)

Today's bioethicists also worry that not enough is known about cloning to ensure that the copy would be as healthy as the original. Dolly, for example, died in 2003, far short of the normal twelve-year lifespan of a typical sheep, apparently because the aging process was somehow speeded up (she was put down after suffering from arthritis and a lung condition usually only seen in much older animals). Would a cloned person be bald and senile before he reached his twenties? Would he suffer from unimagined, catastrophic diseases because of some glitch in the cloning process? Scientists don't know and, for the most part, don't want to find out.

Decades ago, when the technology of cloning was far less advanced, it was a lot easier to believe that cloning a person was inevitable. But as the possibility became more realistic, people became more uneasy with the concept. Around the world, laws are being written to prevent the arrival of the cloned person, yet a few scientists continue to pursue that goal. Time will tell whether lawmakers can continue to stave off the arrival of the human clone.

Prediction: Natural Childbirth Will Be Criminalized

Cloning, ectogenesis, and many other promised medical advances left a lot of people distinctly queasy by the mid-1960s, so much so that lots of parents-to-be sought to turn back the clock. In a nice bit of irony, many bra-burning, liberated women, so forward-looking in other ways, wanted to give birth just like their great-grandmothers did—no beeping machines, no drugs, no sterile delivery rooms. Instead these women, who believed that science had taken all the grandeur out of bringing a life into this world, longed for something more authentic, more natural. That's where the midwife came in.

Back in the days when applying leeches was a cutting-edge medical treatment, the best advice an expectant mother could get came from a midwife, typically an older woman who had given birth herself or who had assisted at a lot of births. This is still the case today in poorer parts of the world where modern medical care is virtually nonexistent.

But in North America, the midwife began yielding to the gynecologist early in the 20th century, with more and more babies born in hospitals, and as a result, this traditional occupation had gone the way of the blacksmith by the 1940s. Midwives fell out of favor largely due to an organized campaign begun by medical doctors around 1900. The physicians argued that far too many babies and mothers were lost under the care of a midwife, who lacked both the equipment and the knowledge to assist during difficult births. Later, natural birth advocates took a much more cynical view, contending that the doctors were really just trying to protect their market from competition.

While midwives were still only employed for a small number of births in the 1960s and '70s, some commentators believed that a panicky medical establishment, fearful that the movement would catch on and a prime source of revenue would be lost, would once again seek to marginalize the midwife—even outlaw her. In 1979, Norma Swenson, who graduated with a master's degree from Harvard's School of Public Health, warned that by now doctors might have successfully lobbied for laws that would make it a crime for women to give birth outside of specially certified medical centers. Pregnant women would be forced to register with a regional medical authority and a computer chip would be implanted in their bodies that would transmit data during every stage of their pregnancy. Women who tried to defy this regimen would have their unborn children taken into protective custody and would be charged with child endangerment.

This may sound far-fetched, until you examine the anecdotal evidence Swenson had compiled. In 1977 the executive director of the American College of Obstetricians and Gynecologists declared that home births constituted child abuse. A woman who refused pain medications and induced labor treatments at Boston City Hospital in 1978 had her newborn placed in a foster home for several months before the woman won a court battle to get the child back. A natural birth advocate, Cindy Duffy, predicted in 1979 that many people like her would be "going to jail in the next twenty years."

What actually had occurred by 2000 was not nearly as dire as all that, but it was still a bit of a mixed bag. Midwives were by then licensed, highly trained professionals (many, in fact, are former OB-GYN nurses), and while they're still used in only a small fraction of U.S. births, the percentage has held steady for three decades. And midwives were even more common in other industrialized nations; in Western Europe they assist in up to 70 percent of all births.

Another variant of natural childbirth, labor in a hospital but without drugs, had also become more popular, even among women without any religious objections like those mentioned earlier. Thousands of women and their coaches take classes in Lamaze and other natural birthing methods each year. And dads, once banished to the waiting room, are now expected to be at the mom's side when the baby (or babies) arrives.

Still, the average American birth, for better or worse, is also more tech-heavy than it has ever been. Cesarean sections are used with growing frequency. Induced labor is now the norm. The menu of treatments and medications used during birth has expanded dramatically. Women who today eschew modern technology during birth may not be the felons of Swenson's worst nightmares, but neither are they typical.

Of course, once you got into this world—whether in a modern hospital or in the parlor of the old ancestral homestead—science wasn't done with you yet. And although the tools at the disposal of the average surgeon today would have astounded their counterparts of a generation ago, they still pale in comparison with what many futurists thought that checkup was going to be like in 2000.

Prediction: There Will Be "Cures" for Obesity and Alcoholism

There once was a time when ministering to the sick was expected to be a much simpler affair than it actually became. In the 19th century, J. E. Erichsen, the surgeon-extraordinary to Queen Victoria and author of the definitive medical text used by doctors during the Civil War, believed surgery had pretty much reached its high-water mark long before 1900. "The abdomen, the chest and the brain will be forever shut from the intrusion of the wise and humane surgeon," he wrote in 1873.

Sixty years later, long after it became clear that Erichsen had perhaps sold his profession a little short, C. C. Furnas, the Yale professor mentioned earlier, predicted that the pharmaceutical industry wouldn't have all that much to do later in the century: "On the medical side of the picture, we may look forward to the day when we will need but few medications but those few will be highly effective." Big Pharma was not expected to be so big after all.

It's difficult to say what Furnas or Erichsen would have made of attempts to tackle humdrum problems such as obesity, the cold, or the toothache. But the midcentury prognosticators who thought that in 2000 we'd all be slim and sniffle free probably could have benefited from Furnas's and Erichsen's cautious outlook.

In 1954 the U.S. military spotted what it considered a disturbing trend: because of vastly improved neonatal care, lean, scrawny babies that before would have died young were surviving into adulthood, producing a generation of wimpy string beans. A survey of fifty thousand G.I.s that year revealed that more than a fifth were tall and skinny, a phenomenon that appeared to be on the upswing. But thanks to less active postwar lifestyles and the fatty, sugary American diet, that trend was short-lived. Within a few years doctors began to notice a different problem: the average American was becoming a lot chubbier.

Doctors and nutritionists soon came up with surefire, commonsense methods to keep our waistlines in check, techniques that but for some particulars have remained unchanged ever since: eat more fruits and vegetables, lay off the fried foods, and take the stairs once in a while. But other researchers worked on various shortcuts, ways that would allow even those of us without willpower to have our cheesecake and eat it, too.

Scientists at Yale in 1966 developed a shot that affected rats' desire for food, "tricking" the animals into believing they were sated. The process could also be adjusted to induce hunger, too—a potential boon in treating the chronically ill, who often have no appetite. (Scientists in the early 21st century had stumbled across a completely natural method for getting the very sick to eat: a toke of marijuana. But in America, strict federal drug laws close off this option for most of the people who could benefit from it.)

Nearly a decade later, Dr. Cleamond Eskelson, at Tuscon's Veterans Administration hospital, developed a treatment that employed a different principle. Eskelson gave lab rats a dose of acetic acid, which slowed the absorption of food into the intestinal tract. With this method a typical person could theoretically consume thousands of calories more than recommended and still be able to squeeze into those hip-hugger jeans. This method also seemed to work for heavy drinkers: the blood-alcohol level of rats fed liquor laced with acetic acid did not go up.

So promising did this line of research appear that in the mid-1960s the futurists Kahn and Wiener strongly believed that safe, effective appetite control would be widely available by the early 21st century. Thirty years ago, the Institute for the Future predicted that the need to diet "[would] become a thing of the past" by 1985, while others thought that the supermarket of today would be stocked with foods containing chemicals that helped us effortlessly maintain ideal body weight. Ways for us to have "that 12th drink" without ill effects were also expected.

Today, of course, there are dozens of so-called dietary supplements on the market designed to help curb appetite, but that's a far cry from what researchers were aiming for back then—a method that would allow us to overindulge without consequences. Most modern diet aids are merely stimulants that do help quell hunger pangs but only by speeding up a person's metabolism (such products would have been well-known to researchers such as Eskelson). Even today, the only way for the alcoholic to get sober is by putting down that shot glass. The safest, most effective way to drop a few pounds continues to be sensible diet and lifestyle choices.

Prediction: A Cure for the Cold

A generation or two ago, researchers were of course also trying to help those whose problems were not self-inflicted. And given medicine's incredible string of successes as of the 1950s, almost no ailment seemed impervious to a cure. Sulfa drugs made childbirth a lot safer than it had ever been before. Polio had been vanquished. Vaccines saved millions from the flu, one of the world's most potent killers. By midcentury, scientists were on their way to wiping out ancient scourges such as smallpox and cholera. If they were having such success with those exotic and complex diseases, then surely the common cold, one of the most widespread infections on the planet, was next.

It did not take creative thinkers long to dream up all the different ways that science might solve this nagging problem of sore throat, runny nose, and cough. In 1956, science writer Victor Cohn imagined a scene at the typical breakfast table, circa 1999: "Timmy Future" is enjoying a bowl of "Super-Mishmosh" cereal loaded with amino acids, proteins, and vitamins that help prevent him from ever getting the sniffles. But if the whole eating-breakfast-cereal thing didn't work out, futurists rationalized that at least one day everyone could be immunized against any bug that came down the pike, from the flu to the then-unforeseen AIDS virus. After all, vaccines were all the rage back then. More than a decade later, Theodore Gordon, a futurist and engineer with Douglas Aircraft who worked on the U.S. space program, believed that soon colds could become so rare that getting one would be occasion for a party: "It might bring a week of nostalgia and evoke fond memories of the time of aspirin and the hot water bottle."

Party or not, Gordon and many other scientists back then made their pronouncements because they assumed that doctors would continue to get better and better at fighting viral infections such as colds.

But this turned out to be a much taller order than they realized. What the midcentury futurists failed to understand was that colds and other viral infections are nature's extreme opportunists: when a medicine is developed that effectively combats one strain, many viruses simply evolve into something that renders the medicine ineffective. The rate at which the viruses evolve makes it difficult if not impossible to design a catch-all vaccination for the common cold.

Today some scientists are worried that if we simply keep developing stronger antiviral medications we will inadvertently help create superbugs that can't even be treated, much less cured. While there are more drastic treatments that can be undertaken for the young and the old, most doctors advocate allowing the immune system to run its course, even if that means feeling lousy for a couple of weeks. (In other words, it's probably time to cancel the keg you just ordered in honor of your next cold.)

Prediction: Cavity Inoculations and Other Painless Dental Techniques

This distinction between treatment and cure was much on the minds of another group of 20th-century medical practitioners: dentists. The vanguard in this profession believed that one day, advanced preventative techniques would give us all perfect teeth without once ever having to succumb to the drill.

By the late 1960s researchers had pinpointed bacteria as one of the main agents that cause cavities. Scientists tried to attack this problem as they would any other infection: with penicillin, which proved to be highly effective in experiments with hamsters. Shots alone arrested tooth decay in generations of the tiny creatures. Doctors quickly recognized that this treatment could not be repeated in people, however; if humans were to receive penicillin constantly, their bodies would soon build an immunity to the drug that would make it useless against the other diseases penicillin fought. Still, scientists for years afterward tried in vain to find a "cavity shot."

A few years later some dentists were experimenting with lasers in the belief that blasting tooth enamel just once could keep your pearly whites pearly white for a lifetime. Another technique, pioneered by R. F. Boehm, a University of Utah engineering professor, employed lasers to painlessly fuse caps onto cavity-prone molars, protecting them forever.

Other dentists—probably those who weren't too keen on advances that ultimately would have cost them their jobs—set themselves the more modest goal of making that stay in the chair a little more palatable. Even with Novocain, getting a tooth extracted was and is an unpleasant experience both for the dentist and the patient, who must sit still while a pair of forceps wrenches the rotted tooth loose.

But some hoped that by now teeth could be jogged free using high-intensity sound waves. During this painless process, which was developed at Drexel University in Philadelphia in the early 1970s, the tooth would be gently loosened from the bone through sound waves before it was manually pulled out.

Even if science ultimately failed to take all of the dread out of dental surgery, others thought it could at least be made more convenient. Thirty years ago a company developed a twenty-eight-pound mobile dentist's office that could be brought to the home or the workplace, because there's nothing better than a trip to the dentist's office in your living room.

Prediction: Pain Relief Will Be as Easy as Pushing a Button

Life for recuperating cancer sufferers was expected to be much better in 2000 than it actually turned out. Analgesics are a class of medications used to control pain, and while they are far more advanced today than they were a generation ago, there are still some drawbacks. The medications often have a host of unpleasant side effects, such as constipation, vomiting, or drowsiness. Some patients also have strong allergic and other types of reactions to their medicines. Many also worry about getting hooked on their painkillers.

Scientists, however, once believed that pain could be conquered using a technique that had no other effect on the sufferer than to remove his or her agony: electrical nerve stimulation, often described as "push-button pain control." It was to be as easy to use as its name suggested.

For reasons that are still not completely understood, administering a bit of electricity seems to interfere with the transmission of pain signals (which are also electrical) through the nervous system. As scientists learned more about human nerve cells, they started experimenting with ways to control the transmission of impulses through the body as a way to treat and cure some patients. In 1965, researchers at Harvard, working with a man suffering from cancer of the larynx, ran electrodes into a region of the man's brain known as the thalamus. After scientists shot a few volts of electricity into this area, the man was able to forgo his conventional pain medications for three months. Around the same time, doctors tried an even less intrusive technique, applying electrical stimulation to "peripheral" nerves in the arms and legs. The futurist Alvin Toffler and others thought that by now the recuperating patient would be walking around with a portable device, tapping a button to deliver a few soothing volts whenever he or she felt the onset of pain.

Today at least a few people benefit from implanted devices to keep their aches and discomfort in check, and the technology in this area is rapidly improving. Still, the primary pain management technique of choice continues to be opiate-style drugs. A modern primer on pain management put out by the American Cancer Society, for instance, goes into great detail about contending with the side effects of medications or how to avoid addiction but contains just a single passing reference to electrical nerve stimulation.

Prediction: It Will Be Possible to Regenerate Body Parts

In the early '70s Robert Becker, a researcher at the Veterans Administration hospital in Syracuse, New York, grew a stump on a rat by applying an electrode in the region where the animal's limb had been amputated. In less than a week, Becker reported, a full complement of bone, muscle, cartilage, and nerve tissue had come back. Though it was far from being a new, functioning limb, Becker was encouraged by the results. The U.S. government—whose veterans' hospitals were filling up with returning soldiers who had lost arms and legs in Vietnam—was also highly interested in Becker's work, so much so that it provided him with a team of fourteen researchers to continue his investigations.

Becker was building on earlier research conducted throughout the century by others who tried, and had at least limited success, in growing fully functioning tissue. In the 1940s a Russian, L. V. Polezhaev, induced some regeneration in frogs simply by poking their stumps with an ordinary needle. In the '50s, a doctor working at Case Western Reserve University in Cleveland regenerated amputated frog tissue by adding more nerve cells. Although he was cautious in his pronouncements, saying in 1973 that any major breakthrough was still decades away, Becker nevertheless thought that "bioelectricity has the potential for changing the climate of medicine as much as antibiotics did." Others shared Becker's optimism, at least to an extent; in 1967 one group of futurists who were devising a chart of potential advances listed regeneration of limbs as an outside possibility by the year 1986.

Electrical regeneration could not only be applied for growing arms and legs, Becker and others believed, but to whole new organs, too, such as hearts, livers, and lungs. This was just one answer to a problem that, ironically, arose due to advances in organ transplantation. There had been steady progress in this field since the first successful skin grafts in the early 1900s, which was followed by the transplant of a kidney in the 1950s, and the holy grail, the first heart transplant, performed by South African surgeon Christiaan Barnard in 1967. Doctors have learned so much about how to use drugs to fool the patient's immune system into accepting a foreign body part that today everything from lungs to penises is transplantable.

·   But now that science had created a demand by perfecting transplantation, what about the supply? Even when the discipline was in its infancy doctors could foresee the time when they would have many patients who could be helped if only suitable tissues from live donors or the recently deceased could be found. That's why many researchers, including Becker, tried to cut the donor out of the equation altogether by working on alternate methods to create a large pool of "spare parts." In the '60s and '70s those who believed science would eventually provide a suitable answer included Dr. Paul Russell of Massachusetts General Hospital, who thought that by using a chemical variation of Becker's technique, organs could one day be grown in the test tube.

Today we still can't grow organs, although research into stem cells, mentioned earlier in this chapter, probably holds out the best hope for regenerating new tissue. Stem cells are unique in that they are "unspecialized," meaning they can grow into almost any other kind of human cell. Should scientists one day learn how to master the process, a clump of stem cells could be tweaked to induce growth of, say, new heart tissue or nerve cells. (Even though stem cell research is at the eye of a political storm in the United States, overlapping as it does with the nation's ongoing abortion debate, doctors in other countries are doing groundbreaking work.) But that day still seems a long way off.

Prediction: Mechanical or Animal Organs Will Fulfill All Transplant Needs

A decade before Becker was performing his regeneration experiments, another group of scientists was looking for other sources of organs, including animals and man-made machines that would be even better than the organs they replaced.

In 1963, a chimpanzee's kidney was implanted in a New Orleans man with the redolently southern name: Jefferson Davis. Davis, a forty-four-year-old dockworker with four children, soon had lots of company: Davis's physician, Dr. Keith Reemtsma of Tulane University, made a dozen other attempts to put chimpanzee kidneys into people, an example of a procedure that came to be known as xenotransplantation. In 1977, Barnard, who had done the first heart transplant, tried to use primate hearts to sustain people while they awaited a suitable donation from a person. Seven years later, the world was captivated by the plight of "Baby Fae," who received a baboon heart when she was just a newborn (the little girl's body ultimately rejected the transplant and she died when she was less than a month old).

Those who didn't believe animals were the future looked forward hopefully to the day when parts could virtually be built to order, and, even more astoundingly, those artificial organs would be deluxe versions of the originals. "[O]rgan technology suggests the possibility of track stars with extra-capacity lungs or hearts; sculptors with a neural device that intensifies sensitivity to texture; *lovers with* sex-intensifying neural machinery [author's emphasis]. In short, we shall no longer implant merely to save a life, but to enhance it—to make possible the achievement of moods, states, conditions or ecstasies that are presently beyond us," futurist Alvin Toffler wrote more than thirty-five years ago. Rattray Taylor, the science writer mentioned earlier, imagined a time when many people would possess bodies with just a few "original" parts—everything else would either be donated or man-made.

Of course, the sad reality today is that the science has not progressed nearly as far as some of the optimists had once hoped. No one today is walking around with a mechanical lung, kidney, or stomach, much less components that would help someone enjoy exponentially better sex, as Toffler once breathlessly described. True, a mechanical heart was invented, but this proved to be less than the earth-shattering breakthrough it was thought to have been when one was successfully implanted in dentist Barney Clark back in 1982. For one thing, the device falls far short of being a perfect replacement (Clark lived just 112 days after his Jarvik-7 was placed in his chest). Also, the heart is a fairly simple organ, essentially just a pump. It's a lot harder to make, say, a mechanical, implantable version of the kidney or the lung, which are both much more intricate. Although work does continue in this field—the prospects for artificial skin, which is used mostly by burn victims, are fairly good—the best parts, and mostly the only ones, are still built by Mother Nature, not us.

Similarly, using animal organs in people has had only limited success—implanting pig valves and clusters of animal cells into human hearts is now fairly routine, but scientists still haven't figured out a reliable way to make a person's immune system accept a complete animal organ. One of Dr. Reemtsma's patients who received a primate kidney in the early '60s did survive for about nine months, but for the most part the dozens of other xenotransplants performed during the last forty years ended like Baby Fae's—in immediate, tragic failure. Research is nevertheless still being conducted to find ways to make more of a beast's parts work in a person. Scientists have speculated that pigs, whose organs are very similar to ours, could be genetically modified so that their immune responses are closer to a human's. In theory, the immune system of the recipient would then "recognize" the pig part and it would not be rejected. Developments like this, however, are still a very long way off.

Prediction: Life Begins at 140 and Other Bad Calls About Aging

The scientists who tried to discover ways to make organs available to all who needed them were, in a sense, merely attacking the branches of the problem, not the root. What if we could prevent organs from ever deteriorating at all, and discover a way to halt the slow, steady decline of aging itself?

Those with access to modern medical care now live a lot longer than people typically used to. Thanks to advances in orthopedic surgery and the treatment of heart disease, for many people who are careful about their health seventy is the new fifty. It used to be that living to one hundred was a goal attained by so few that someone's first three-digit birthday made the news. Now there are forty thousand centenarians in the United States and people at least one hundred years old are the fastest-growing segment of the population (the segment between eighty-five and one hundred is the second-fastest). In 1997, Jeanne Calment of France died at the age of 122—the longest anyone is known to have lived.

None of this suggests that science is any closer to cracking the aging riddle now than it was a century ago, about the time when gerentology, the study of aging, first emerged. Average life expectancy in America today may be much higher than in 1900, but that's partly because almost no one dies of cholera in infancy anymore. Getting a heart attack at forty-five is no longer the death sentence it once was. With millions of people in the industrialized world now holding office jobs, there are far fewer fatal workplace accidents, too. But these developments have had no effect on life span itself—the longest a well-maintained human body can last today is pretty much the same as it was a thousand years ago (the Italian painter Titian, for example, may have been ninety-nine when he died in the 16th century). The only difference now is that advances in medical care and more widespread wealth have greatly increased one's chances of living beyond the century mark.

Decades ago some scientists recognized this and began trying to untangle the question of aging by looking at it as a cellular, chemical, and genetic process. Gerontologists back then believed that aging could potentially be reversed, that years—even decades—could be added to the maximum life span of one-hundred-odd years, or that treatments would be discovered that would rejuvenate us in our old age. With the onset of 20th-century medical advances, many began dreaming of the time when all of us could still squeeze out that tenth chin-up well into our eighties. Consider this lovely, hopeful outlook for the early 21st century, written in 1925: "And all the world will be young. The advances of medicine and surgery will be such that most of the ailments and limitations of old age will have been eliminated. Life will be prolonged at its maximum of efficiency until death comes like sunset, and is met without pain or reluctance."

Scientists in the '60s thought they had found a way to literally breathe life into this poetical vision, at least as far as a person's mental capacity was concerned. Memory, reflex response, and the ease with which we learn new things all decline with age, but in 1967 researchers thought that a few puffs of specially treated air could sharpen the minds of seniors almost instantly. Old laboratory rats exposed to ionized (that is, electrically charged) air were able to complete a maze in just under twelve minutes, while the whippersnappers they were competing against—a control group that had only breathed ordinary air—took nearly four times as long. Results of this experiment confirmed those from similar tests done in 1959.

Charged air was but one of an array of proposed rejuvenators, at least a few of which were supposed to be available by now. Dr. L. H. McDaniel, who spoke before the American Medical Association's 104th annual convention in 1955, predicted that by 1999, women would be able to stay young, beautiful, and shapely for their entire lives with just a few hormone shots. Other scientists hoped that extended life could be achieved in stages of renewal: one treatment would keep you going until the next antiaging treatment came along.

In the late '70s, Denham Harman at the University of Nebraska discovered that several compounds, including vitamin E and a food preservative called intoquin, helped increase life span in mice by 50 percent. The chemicals Harman used were antioxidants that helped offset the effects of free radicals, molecular fragments that disintegrate tissues as they knock about the body. Another antioxidant, centrophenoxine, was used, like the ionized air treatments, to restore the mental faculties of senile patients. Famed gerontology researcher Roy Walford and others experimented with a technique to lower body temperature by a degree or two while people slept, a method that some believed could add thirty years to a person's life—or even double it.

But toying with body temperature was just a sideline for Walford, who believed in prolonging life in other ways. Walford's biggest claim to fame was the ultralow-calorie starvation diet, a practice that has been promoted in one form or another since a famous series of experiments eighty years ago involving, yet again, rats. In 1927, Dr. Clive McCay, who observed that the aging process is very slow in vertebrates, such as sea turtles, that continue to grow throughout their life, deliberately underfed rats while they were still maturing. This had the effect of slowing down their development, so the rats were still at the juvenile stage when they were one thousand days old. (In human terms, this would be like a seventy-year-old just entering puberty.) At this point, the rats' caloric intake was increased to normal levels and they lived for another five hundred days—or half as long as they normally would. Even more exciting, the Methuselah rodents were as frisky as teenagers. If the same results could be achieved in people, we would enjoy eight or nine decades of youthful vigor until we started aging. The golden years would be considered anything past 110 or so.

The tricky part, at least in the rats, was trying to figure out when normal feedings should begin. In a variation of McCay's original experiment, for example, half the rats that were underfed for 1,150 days were unable to resume normal growth. Walford, an accomplished scientist and a bit of an eccentric (he spent sabbaticals walking across India in a loincloth, meeting with yogis), thought he had the answer. He was the author of *The 120-Year Diet*, in which he promoted eating far fewer calories than was generally recommended, a regimen he followed for decades. Walford consumed only about 1,600 calories a day—about half what's considered healthy for an active man—after a series of experiments with sparse diets he conducted in the 1960s doubled the life span of mice. Walford's program caught on, even though it didn't seem to work for the man who pioneered it: he died of Lou Gehrig's disease in 2004 at the respectable but by no means extraordinary age of seventy-nine.

Walford was fairly modest in his aims—after all, he was merely promising that more of us could get our natural "maximum allowance" of longevity. Other researchers, however, were swinging for the fences. Forty years ago noted gerontologist Robert Prehoda was virtually positive that a true antiaging elixir would be found by the 1990s. Prehoda's starting point in his quest was to explore the evolutionary usefulness of growing old. Nature seems to want us to live just long enough to raise our offspring. That job finished, the timer goes off on our biological clocks and we start falling apart. By dropping dead once we've outlived our usefulness, more food and other resources are available for the up-and-coming—an added bonus from the species' point of view. The task was to "rewind" the clock, a problem that Prehoda believed had to be attacked on many fronts.

Prehoda cited the work of Johan Bjorksten, who was exploring how tissues in the body break down over time due to a chemical process called cross-linking. Prehoda believed that a special diet containing a specific soil protein could counteract this effect. Another group of scientists was working on artificial viruses that could be implanted to slow or halt genetic mutations that Prehoda and others thought also played a role in aging. Germs were another culprit that Prehoda thought would be dealt with by those seeking a splash in the fountain of youth. Prehoda was confident that if the funding were made available into these lines of research, people would be living to two hundred—at the very least—by the early 21st century. A cautiously optimistic Prehoda also thought that, unless some unknown biological barrier was discovered, we might last five hundred or even a thousand years!

Of course, there was some downside to living as long as a redwood tree. Would a woman who wed at twenty-five still feel that special something for her husband after three hundred years of marriage? And Christmas could break the bank for people expected to buy presents for their great-great-great-great-grandchildren (with a few centuries of life, a person could see three or four thousand of his descendants).

Other possibilities, meanwhile, were downright creepy. Many skeptics of the gerentologists' wildest claims were quick to bring up the Greek legend of Tithonus, who was granted eternal life by Zeus—but not eternal youth. His lover, the goddess Eos, looked on in horror as Tithonus became more decrepit while his heart refused to stop beating. What if science only got good at keeping us out of the grave and nothing more?

Equally disturbing was the possibility of living for several decades—without a body. In the '60s doctors had already succeeded in keeping the isolated brains of dogs and monkeys alive. One physician at the time noted that sustaining a human head would be even easier.

By the dawn of the 21st century it was clear that Prehoda, McCay, Walford, and some of the rest were all right—for what that's worth. Each seemed to be contributing a piece of the puzzle behind aging, but how it all fits together and how we can prevent it from occurring, no one can yet say. In fact, the more we learn about aging, the less we seem to know. For example, until recently it had been widely assumed that life span has a genetic component, an assumption made not just by scientists but laymen, too (it's not uncommon to hear someone brag about a long-lived grandparent, as if that were proof that longevity "runs in the family").

But in 2006 scientists in Europe released the results of a massive study that upended this long-held belief. Looking at the life spans of thousands of twins born between 1870 and 1910, the researchers discovered that siblings identical in every other way still didn't enjoy anything like the same amount of time in this world—the twins died an average of ten years apart. Other studies show that while your parents' height is a very good predictor of how tall you'll be, there is little correlation between a mother's or father's longevity and their children's. Genetics, it turns out, doesn't seem to have much of a role in determining life span after all.

Scientists today are still hopeful that true longevity can be achieved, and some are still predicting that the 130-year lifespan, or the equivalent, will be coming in the near future. But given the track record, most serious researchers in the field are a little less optimistic than they had been before.

Prediction: You're as Cold as Ice

In 1962, a five-year-old boy named Roger Arnsten fell into a freezing Norwegian river and spent twenty-two minutes underwater. He was dead for more than two hours. His body temperature plummeted to 75 degrees, more than 20 degrees below normal. Just a few months later, little Roger had made an almost complete recovery.

How did Roger survive? His terrifying plunge effectively shocked his system to a halt: his heart stopped beating, his lungs stopped breathing, and his oxygen-starved brain shut off. But, interestingly, the same cold that killed him also slowed down the decay of his tissues. When most people die, their brain cells immediately begin disintegrating. Luckily for Roger, though, his brain cells were quick-frozen, like cod caught fresh at sea. When Roger's doctor, Tone Dahl Kvittingen, restarted his heartbeat, Roger's still-intact brain simply switched back on.

What if this effect could be routinely duplicated? What if those at death's door today could be dipped in an icy bath of liquid nitrogen and kept frozen until scientists figured out a way to cure them? What if quadriplegics were given the opportunity to go under now and wake up with a new, fully functioning body in 2050?

To one man, Robert C. W. Ettinger, those possibilities were so tantalizing that he believed society was morally obligated to realize them. His 1966 book, *The Prospect of Immortality*, is an intricate argument in defense of cryonic freezing that explored its legal, scientific, and even religious implications. Ettinger also predicted that by now a person's death would be regarded merely as the interlude before he or she got a second crack at a new, better life.

Ettinger's basic argument was that when tissues are stored near absolute zero, the decay process stops. Even though the science of the 1960s was not advanced enough to revive a human being who had been put in long-term deep freeze, enough was known about cryonics back then to preserve the body for thousands of years—giving science ample time to figure out a way to bring back all of those iced people. Why,

Ettinger wondered, shouldn't the world institute a massive freezer program immediately for that reason alone?

Ettinger also predicted that once a simple freezer program had begun, scientists could continue refining the process, and that work was in fact under way in the 1960s. Doctors had already successfully inseminated women with sperm that had been frozen for years (something similar is routinely done today at thousands of sperm and ova banks). Certain insects and simple animals such as clams were frozen and revived hours or days later. So too were some animal organs. Scientists had already discovered that certain cold-weather bugs produced their own, natural antifreeze, leading some to wonder whether a similar substance could be developed for use in people.

But in a world where the cemetery or crematorium was no longer the final destination, what would death mean, legally? Ettinger thought that for but a few cases, not much. Unless a body were completely destroyed, no one could any longer be declared "dead" in the way the term has been used since time immemorial. Instead doctors would be required to ice down cadavers immediately—failing to do so would be the equivalent of denying potentially lifesaving medical care. Even if just a bit of a body still remained, that too would have to go in the freezer (under the assumption that by about 2100 or so our descendants would have figured out a way to regenerate a person from just a few cells). There were other, thorny questions that needed to be worked out, too. Was someone free to remarry even though their dear departed spouse might one day become undeparted?

Even with the legal niceties squared away, another big question still remained: Who was going to pay for all this? Ettinger suggested that in America, the Social Security Administration could be revamped so that not only an income would be provided in one's later years, but also a constant supply of liquid nitrogen in one's, um, post-later years. And since no one would really be dead anymore, their assets would continue to be their property. A few thousand dollars squirreled away when the freezer door was slammed shut would be a very tidy sum after you factor in a few centuries' worth of compound interest.

Ettinger even devised some clever arguments for anyone who had religious objections. To those who would say that God "intends" for us to die, Ettinger pointed out that there is nothing unholy about the car, even though we're not born with wheels or the ability to go 100 miles per hour. Another concern involved the nature of one's immortal soul. In the Christian tradition, it's believed that a single spirit is wedded to an individual body; when the body dies, the soul departs. Bringing someone back from the dead, therefore, would be like trying to drag them back down from Heaven (or hauling them up from Hell, as the case may be).

To this Etttinger points out that in other religious traditions, the concept of soul is radically different. Hindus believe that an individual experiences myriad reincarnations, so returning to their thawed, revived body after a few hundred years presents no theological conundrum. In Shinto and some other faiths, meanwhile, there's a belief in one all-encompassing spirit, not billions each assigned to a particular lump of flesh. Given this uncertainty, Ettinger argued, it was best to err on the side of verifiable science.

While opinion about cryonic freezing was decidedly more mixed among many mainstream scientists, at least a few laymen were famously sold on the concept. When baseball great Ted Williams died in 2002, his son John Henry caused a furor when he had the famed Red Sox slugger's body transferred to a cryonic warehouse in Arizona. According to others who were close to Williams, this was not consistent with his wishes; Williams apparently wanted to be cremated. Some raised the rather macabre possibility that Williams's son hoped to sell off his father's DNA. (Contrary to popular belief, Walt Disney is not among the few thousands whose bodies are now being kept in cold storage. The father of Mickey, Minnie, and Goofy has his ashes buried in a Southern California cemetery.)

Would it have been possible, in any event, to one day again see Williams, or Williams 2.0, knock one out of the park? Many scientists doubt it. Even if Williams's body could be restored, many believe his essence—his mind—is irretrievably gone. The interplay of electrical signals and neurochemicals that make us us cannot be re-created after an extended period of brain death, whether we're frozen or not. Until someone is successfully revived—and that's a big if—a lot of scientists will remain skeptical.

Cryonic advocates argue, however, that while short-term memories disappear with clinical death, long-term memories recorded in the brain's very structure preserve all our personality traits, knowledge, and habits. Preserve the brain, then, and you're preserving the person.

Cryonic freezing wasn't the only way we were going to tack on a few years, however. Twentieth-century researchers studying the mechanics of animal hibernation believed that a similar process could be devised for people. And hibernation—or its close cousin, suspended animation—held out the promise for hyperlong life.

Suspended animation continues to turn up in the pages of science fiction, where it enables intrepid space travelers to survive that thousand-year journey to Omicron-6. More realistically, NASA scientists in the 1960s were interested in the concept for a much shorter jaunt to neighboring Mars. If all of the body's processes could be slowed down (as bears, some hamsters, and hummingbirds are all able to do), a manned mission to the Red Planet would require a lot less food and oxygen, cutting down the vehicle's weight by thousands of pounds. The astronauts would also have experienced less wear and tear on their bodies. Those months in hibernation could, some scientists thought, pay off as perhaps decades of additional life.

Experiments conducted in the mid-1960s at the University of Missouri demonstrated that hibernation is a genetically inherited characteristic. This means that the hibernation gene could, in theory, be isolated and tailored for use in humans. Some speculated that a shot could be developed that, if taken nightly, could reduce heart rate to 10 percent of normal. In theory, a one-hundred-year-old man could have the coronary capacity of a thirty-year-old after a lifetime of such treatments. Slowing down metabolism, Dr. Robert de Ropp speculated, could add a quarter century to our life spans.

Futurists well into the 1970s gave suspended animation at least a remote chance of becoming reality by the early 21st century.

Prediction: I Can Read Your Mind—for Real

Mention ESP and many people might automatically make an association with lurid tabloid newspaper headlines and "researchers" boasting mail-order Ph.D.s. They might not associate such work with one of America's most prestigious colleges, however.

But until early 2007, when it closed down for lack of funding, a cramped laboratory at Princeton University was home to what was considered to be the vanguard research institution examining extrasensory perception, a catchall term that describes everything from an ability to read minds to actually influencing the physical world through the power of thought alone. Not that the Ivy League school was sorry to see the lab go: when it shut down due to a lack of funding, Princeton didn't even release an official statement.

But where mainstream academia was disdainful of the work being done at the Princeton Engineering Anomalies Research laboratory (PEAR), there were a few heavy-hitting adherents, such as PEAR's founder, Princeton physics professor Robert Jahn; James McDonnell, who helped start the McDonnell Aircraft Corporation; and Laurance Rockefeller, like McDonnell a generous donor. Half a century ago, when scientists were first beginning to grasp how the mind worked, their support for ESP research would not have seemed quite as strange as it does today.

In the mid-1960s, scientists at the University of Edinburgh, the University of Virginia, and the University of Southern California, among others, were all conducting various ESP-style experiments. In one experiment, a "sender" in one room was shown a provocative image—the assassination of President Kennedy, a naked woman—while a "receiver" in another room was asked to record whatever thoughts or impressions popped into his or her head. Twenty-eight out of thirty-two receivers reported impressions appropriate to the material shown the sender (that is, horror at the assassination photo), while one

receiver reported an impression of palm trees and Honolulu after the sender had been shown an image of a tropical beach.

Twenty years later, as even more had been learned about
the brain, eminent (though controversial) scientists such
as José Delgado, who had taught at Yale, predicted that by
about 1980 or so, techniques would be developed to foster
"nonsensory communication" between human brains. In the
1990s, Brian Josephson, who won a Nobel Prize in physics,
cited other experiments showing that thoughts can be communicated successfully about a third of the time—more frequently than random chance alone would dictate. In the early
1980s, he believed that experiments showing a person's ability
to bend a spoon with his mind would, within a decade, be
considered "highly significant" by the mainstream scientific
community.

An offshoot of ESP and parapsychology involved so-called
lucid dreams, experiences so vivid the slumbering person is
actually aware he or she is in a dream state. In the early 1980s,
scientists at Stanford University's Sleep Research Center en-
listed a team of "oneirauts" (dream sailors) to study their own
sleeping sojourns and even manipulate them. Stephen LaB-
erge, who was leading the research, believed that inducing
lucid dreams would eventually lead to methods of analyzing
them, giving people unprecedented ability to examine their
own deep-seated fears and anxieties. Another researcher was
hoping to patent a machine that would jolt the dreamer to
awareness at the onset of a powerful dream, giving the individual the ability to step into his own dream world and control it. Not only would you be able to stop and speak to that
unseen demon chasing you down a dark alley, you could also
find out why it was chasing you in the first place.

Today, ESP, dream manipulation, and study into parapsychology generally just barely hang on at the margins of respectable academic research (as of 2002, about a dozen Ph.D.s in parapsychology held teaching posts at British universities). For the most part, though, belief in ESP is considered among most educated people to be on a par with fortune-telling and palmistry. Why, given the seemingly promising experimental results and the support of at least a few highly respected scientists, had this area of research largely foundered well before the turn of the century?

Much of the blame can probably be placed with some of the very people who wanted most to believe that the powers of the mind could shape the physical world around us. By the late 1960s, parapsychology had entered the pantheon of loopy New Age fads, an association that undoubtedly scared off many legitimate researchers. As Josephson noted a few years ago, respectable scientific journals won't touch parapsychology and ESP. Because of this, Josephson believes, no one in academia knows about the groundbreaking—and verifiable— work being done in parapsychology.

Prediction: Acid Will Be an Okay Trip (Trust Me)

When LSD was discovered by Swiss chemist Albert Hofmann in 1938, there would still be another twenty years until the drug, derived from a grain fungus, would enjoy its moment in the sun. In that year, Hofmann accidentally absorbed the drug through his skin, thereby becoming one of the first people ever to experience an LSD "trip," an hallucinatory mind-state marked by a heightened awareness of colors and sounds. (He was hardly the first to experience a mind-altering chemical high, of course—for thousands of years indigenous peoples had been using mushrooms, peyote, and other plants in religious ceremonies for that same purpose.) Did Hofmann regret unleashing his discovery on the world? Hardly: "[LSD] gave me an inner joy, an open-mindedness, a gratefulness and an internal sensitivity for the miracles of creation," he told a crowd gathered for his hundredth birthday in 2006.

If Hofmann was the John Locke of the psychedelic movement, however, Timothy Leary was its Thomas Jefferson, the radical who tried (ultimately unsuccessfully) to upset the apple cart in the United States. A Harvard University professor who as a student had been kicked out of West Point, Leary first tripped on mushrooms containing the hallucinogen psilocybin in 1960 before trying Hofmann's creation in 1962. "It was the most shattering experience, of my life," Leary wrote.

Leary would go on to spend the next several years advocating the use of psychedelic drugs as a way to enhance perception, enabling trippers to derive core truths about themselves and the world around them supposedly hidden from the workings of the conscious, intellectual mind. Leary, who died in 1996, is widely regarded today as a lone, mad prophet, but in fact he was hardly alone in predicting broader applications of the drug beyond recreational usage.

In June 1965, scientists in San Diego began a three-year experiment to treat a group of male alcoholics using mild dosages of LSD, under the theory that the drug could alter the men's perspective so they could see that liquor was ruining their lives. The subjects—a typical cross section of people including young single men and grandfathers, blue-collar workers and lawyers, and even an officer in the United States Navy—reported reactions familiar to anyone who has ever indulged. One said he saw "the pure light of the void"; another thought he was at the North Pole. Several compared their trip to a religious awakening and seemed to experience the desired effect, promising to reform their ways and atoning for the hurt they'd caused their loved ones.

Others believed that otherwise healthy people, not just alcoholics, could also benefit from the askew glance of the universe that LSD promised. Hugh Downs, a longtime TV personality who appeared on *20/20* and other shows, wrote in 1967: "We may yet find in [LSD] much to comfort and enhance the individual human, and we may find the key to closing the gap between the runaway pace of our technological progress and the snail's pace of our social change." If soulless modernity was the disease, Leary, Downs, and others thought, LSD might be the cure.

It wasn't just therapists and the peace-and-love crowd that were interested in LSD, however. Starting in the 1950s, the Central Intelligence Agency began slipping doses of acid to unwitting victims under a project code-named MK-ULTRA, which was made public during hearings in the mid-1970s overseen by Senator Ted Kennedy. Although the CIA had destroyed many of the files on the program long before Kennedy's inquiry, some speculated that the CIA had hoped to create *Manchurian Candidate*-style assassins, a project that was supposedly abandoned when the affected subjects proved to be too erratic to attempt hits against high-profile targets. The revelations led to new regulations governing the CIA's actions and also clarifying rules requiring subjects' consent before using them in medical research.

LSD and the very concept of using chemical agents to heighten consciousness had already fallen into disrepute long before MK-ULTRA was made public, however. In the 1960s,

Leary helped link acid conspicuously with the counterculture, which aimed not merely to "enhance the individual human," as Downs put it, but to overturn what the hippies considered to be a morally bankrupt society. Leary encouraged young people to "tune in and drop out," to use the drugs as a path toward a better world where school, career, and family didn't matter so much. But to the people who get to write laws defining what is a narcotic substance, those things mattered a great deal.

And even most adherents of the drug recognized that it was best used carefully, and by some people not at all. Some found the effects so disorienting that they suffered permanent psychological damage; for others LSD induced a state of catatonia. And as with any street drug, impure, unsafe substitutes were soon widely distributed as well, leading to more than a few deaths. By 1970, most countries had outlawed LSD use.

Although LSD's moment came and went, the scientific community and the general public did not give up on the idea of using drugs to supercharge our brains. Alongside the LSD researchers were scientists who thought that one day, learning a new skill—or even tacking on a few dozen IQ points—would simply be a matter of taking a pill.

Prediction: Genetic Engineering Will Make Us Superhuman

Back in the mid-1960s, *Life* magazine tried to provide America a straightforward explanation of advances in the arcane science of genetic engineering. In the near future, the article proclaimed, scientists would be able to tailor-make a baby according to the specifications chosen by the prospective parents. Eye color, height, athletic ability, intelligence, temperament—all could be guaranteed in what some described as a baby supermarket. Reflecting the era's less skeptical attitude toward science, the article presented this possibility as just one more modern convenience, another way man was one-upping Mother Nature for his own benefit.

The article drew a few snickers, but not many. The study of genetics by then had seemed to recast the whole nature-versus-nurture debate squarely in favor of the former. Extensive studies of identical twins raised in separate environments revealed remarkable similarities in such basic qualities as intelligence and height. Twins who had not known each other growing up sometimes even gravitated toward the same occupations—clear indicators that biological factors are at work in influencing a person's makeup. (It's telling that Nazi scientists, ever on the lookout for evidence of inherent racial differences, were keenly interested in the study of twins.) But until the 1950s, scientists were limited to studying the end result of genetic determination, wholly ignorant of how the process worked at the molecular level. Then came Watson and Crick.

In 1953 James Watson, an American, and Francis Crick, a Brit, divined the double-helix structure of deoxyribonucleic acid—DNA—the self-reproducing molecule that carries the genes for nearly every known form of life, including people. Once that was understood, scientists could then examine the genes that make up an individual's genome. This opened up two related possibilities: first, science could determine which genes cause which characteristics (including heritable diseases), and second, scientists could manipulate these genes so as to produce a specific result.

In the 1960s, scientists knew the general outlines of the task and, crucially, knew what they had to do to achieve it. They knew that ever more powerful computers would be coming along to aid the effort, so biologists have spent decades patiently waiting for the programmers and chip designers to deliver up faster and faster machines. Most predictions about where this was to lead us by now have only missed the mark because the technology hasn't advanced quite that far yet or because legal strictures and ethics guidelines have (at least so far) impeded research.

Of the areas where genetic researchers can proceed without any ethical concerns, the treatment of congenital disease tops the list, and any gap between prediction and reality can be chalked up to a mere lack of knowledge. Scientists have not yet found a way to "deactivate" genes that may cause allergies, obesity, and arthritis—as biophysicist Robert Sinsheimer predicted in February 1966—but those problems could very well be tackled within the lifetimes of many people who are reading this book. In fact, identifying some disease-carrying genes—the first step toward finding a cure—has already been accomplished, leading to calls for legislation today that would keep a person's genetic makeup confidential from medical insurers. (The fear is that insurance companies may refuse coverage for people with a predisposition for certain diseases.)

But what about selecting for desirable attributes, such as intelligence, athletic ability, and even good looks? Again, several scientists a generation ago were sure we would be there by now. Indeed, many commentators, including Gordon Rattray Taylor, back then feared that a wealthy, educated elite would demand the best possible genetic inheritance for their offspring that money could buy, whether others could afford the same treatments or not. And once that elite came to maturity, wouldn't they try to restrict access to quality genetic material to their own caste, thus ensuring their comparative advantage? Another thinker, Pierre Auger, worried that genetically modified human beings might come to be considered a different—and more advanced—species, one that has no natural affinity for its primitive forebears.

Ethicists have not yet had to confront these possibilities. Scientists are sure that attributes such as intelligence and of course looks are largely heritable, but they also realize that these highly complex traits involve the interplay of lots of genes. In fact, the "higher-order" attributes all involve a complex ballet of many different chemical components. Sorting that out will probably take quite some time yet.

Other far-out possibilities never came to pass, either, although they might have been easier to achieve. Scientists like Jean Rostand speculated decades ago that, using genetic manipulation, human beings could be endowed with green skin or gills by isolating the genes in plants and fish that produce those traits and splicing them into human DNA. Since then scientists have created numerous such genetic hybrids, mostly involving crops and insects such as the fruit fly. While humans—and vertebrates in general—would be a challenge to work with, the technical obstacles would not be insurmountable.

But even in genetic science, where the moral boundaries are still not perfectly defined, endowing a human being with a wholly nonhuman trait is still taboo.

As technology advances, however, will that ethical sensibility continue to prevent us from turning the human race into something unrecognizable, even horrific? Some observers aren't so sure. Already there are parents who have used technology to determine the gender of their embryos and to select either males or females. Some doctors claim that this is merely a service meant to help families achieve a desired "balance," but others worry that this is the first step toward the baby supermarket described by *Life* magazine more than forty years ago. At the same time, billions of dollars are being spent and the best brains in the world are dedicating themselves to genetic research, virtually guaranteeing that ever thornier questions are going to arise in the next few years.

Chapter Two

GETTING THERE

The hugely popular 1985 film *Back to the Future* ends with Doc Brown picking up Marty McFly in his tricked-out, time-traveling DeLorean.

"Roads?" he tells McFly. "Where we're going we don't need roads." Doc then pushes a button and his DeLorean ascends, zipping off over the horizon and toward the next sequel in the franchise.

In the movie, Doc Brown had just returned from the year 2015, and while that date is still a few years off, it's a safe bet that yes, we will in fact still be needing roads by then. But that's not what the futurists for the past hundred years have been telling us. In the early 20th century, when dirigible flight was first being developed, visionaries believed that the skies over modern cities would be chock full of stately, slow-moving zeppelins, a subway system in the heavens. When airplanes came on the scene a few years later, the covers of pop-science magazines started depicting suburban Dad hovering over his driveway in his sky car, tipping his fedora to the little woman as he headed off to a day at the office. The makers of TV shows and movies got in on the act, too. In the '60s-era cartoon *The Jetsons*, working stiff George Jetson got stuck in traffic jams a few thousand feet up. Along with the *Back to the Future* movies, films such as *Blade Runner, The Fifth Element,* and a host of others featured flying cars, too.

Almost nothing says "future" quite like the flying car, and that's probably because it seems like something so very close to being realized—after all, we've had cars for a long time and we've had planes for a long time; how hard could it be to combine the two? And partly it's because the idea of taking to the skies to get to the grocery store or to drop the kids at school is simply such a neat idea. Who hasn't been stuck in a traffic jam and imagined how cool it would be just to skirt above those long, fuming lines of cars, to get where one needed to go as fast as a bird could?

So how come we're still getting from point A to point B the way our grandfathers and great-grandfathers did? It's not for lack of ingenuity—sky cars that really work have been built, and are getting better all the time. No, the problem seems to be that it would take a lot more than a good prototype to change the way millions of us get around. The transportation infrastructure—the thousands of miles of roadways and railways, the airports, the subway lines—is like the country's nervous system. Tearing it up and reconfiguring it for flying DeLoreans would cost trillions of dollars (the country would need thousands more air traffic controllers, for starters).

The need to completely rejigger America's transport system has so far doomed other alternatives to the traditional automobile. From the 1930s onward, for instance, a lot of people believed that today we would no longer be driving our cars—instead they would be driving us, cruising along "smart" highways that communicated electronically with our trucks and compacts and getting us where we needed to go with just the push of a few buttons. Mass transit was also supposed to get a whole lot smarter and become so convenient that by now everyone would be able to go virtually door to door on the local bus or train system. Ironically, America was a lot closer to having universal mass transit a century ago than it is today—many cities and towns were served by extensive trolley systems, including communities in Southern California, now the car capital of the world.

The trolleys were largely killed off by the automobile companies, who bought them up and gutted them so as to eliminate competition. And a lot of people also blame America's Big Three automakers for undermining the electric car. The first ones were built around 1900, and several times since then—including the present day—the battery-powered, plug-in auto was seemingly poised to dethrone the gas guzzler. But the story of the electric car's numerous failed comebacks just goes to show how hard it is to change an entire national transportation system. When it comes to getting around, we Americans prefer to stick with the tried and true, even if there are other ways to go that are less expensive, better for the environment, and would help get us out of traffic jams.

Put another way, we would just rather repatch and expand the system we have now than take a stab at creating something altogether new. The wide-eyed dreamers of fifty and a hundred years ago never appreciated just how hard it is to get us to change our ways. It looks like it will still be a while yet before we get those flying DeLoreans.

Prediction: There Will Never Be an Affordable Car

When Henry Ford founded his company in 1903, most of the people toiling away on his assembly lines could not afford to buy the cars they were helping build. Back then a car was considered the most luxurious of luxury goods, toys for the Champagne-and-caviar set, not the appliance for the masses it is today. Cars back then were handcrafted, containing fancy styling flourishes that drove the price up to fifty thousand dollars or more, in today's dollars. Before Ford, many of the people buying cars planned to be driven in them by their chauffeurs. Ford, however, envisioned a machine that would be used by farmers and factory workers, something even the chauffeurs themselves could afford. More than a few observers thought Ford was crazy.

"Affluent persons who have got tired navigating Long Island Sound and sailing up and down the Atlantic Coast in yachts find novelty and pleasure in yachting on shore," according to the August 22, 1903, edition of *Harper's Weekly.* A few years earlier, in 1899, the *Literary Digest* predicted that while its price "may fall in the future," the "horseless carriage" will "never come into as common use as the bicycle." Years later, Woodrow Wilson, who was then president of Princeton University, beseeched students to avoid driving their cars in front of the far less wealthy townspeople; he believed that the common man would never have the means to buy a car of his own.

Ford, of course, proved Wilson wrong, by perfecting mass-production techniques that did in fact bring the price down. Ford also realized that in order for his company to be a success, average wages would have to come up, and he practiced what he preached, paying his own workers five dollars a day— not a bad salary back in 1914. Ford's decision to pay such a princely sum scandalized his fellow plutocrats, who thought that the working class would be spoiled lazy if they actually had disposable income. Ford, meanwhile, went on to become one of the world's first billionaire capitalists, thanks to the fact that millions of people were buying Model Ts.

Prediction: The Electric Car Keeps Going . . . and Going

Today, cars powered by electricity or fuel cells are considered an example of hip, cutting-edge technology, a modern green solution to problems created by our parents and grandparents. What a lot of people don't realize is that the electric car was also the flavor of the month more than one hundred years ago. For a very long time, we've been able to build the electric car, and lots of people have been saying we should build the electric car. We just haven't—at least not on a scale large enough to make a difference.

The electric car was, despite its shortcomings, a viable competitor in the market during the car industry's early days. Their batteries were heavy and could only power the cars for a limited range before they needed to be recharged, but for short commutes within small towns, this was not necessarily a problem; in the early part of the 20th century short trips were the only kind most people were making. Around 1910, when electric car production peaked before entering a steep, decades-long decline, there were almost no paved interstate highways, so car travel was limited to jaunts within populated areas, which is where the only good roads were.

Among the first people to bet—and lose—on the future of the electric car was none other than Thomas Edison, who in 1910 declared, "The nickel-iron battery will put . . . the gasoline powered buggies out of business in no time."

As America's highways improved and car trips became longer, however, the electric car was left behind in the oily blue exhaust of the internal combustion engine, which was both more convenient and more powerful. Ford and the other major car manufacturers had common cause with the moguls of America's burgeoning oil industry in promoting a petroleum future. Big Oil and the Big Three had a lot invested in the status quo, so large-scale production of electrics had petered out by the 1930s and wouldn't be tried again for another forty years.

Although Detroit never really put its heart into the effort until it was forced to do so by the oil crises of the 1970s (and even then only reluctantly), some engineers in the 1960s independently began to promote cars powered by cleaner, quieter alternatives to gasoline motors as a way to address the pollution that was by then befouling many major American cities.

Advocates didn't delude themselves about the electric vehicle's appeal. No one believed that the boxy little machines would ever hold their own against the powerful, eight-cylinder behemoths that Detroit was producing at the time—at least as far as highway driving was concerned. But enthusiasts did see another, potentially large role for the electric car—one based on the fact that, surprisingly, the way people used their cars had not changed all that much from Henry Ford's day.

In 1969, the U.S. government released a study showing that about half of all daily car trips were under five miles in length, and 95 percent were under thirty miles in length; just under 2 percent of all journeys were for fifty miles or more. For the short jaunts that constitute most driving, a battery-powered car can be operated much more efficiently than its internal-combustion counterpart (the same conclusion that

Thomas Edison must have drawn sixty years earlier). Although they would still need to be recharged (most models would have required the cars to be plugged into a wall socket overnight), the batteries of the time could easily hold enough juice to power a car through a day of suburban errands. Many experts believed that, in an era when more and more families were buying second and even third cars, one of those vehicles could be an electric.

Naturally, power companies were quite enthusiastic. "Utilities Rooting for Electric Cars," proclaimed a 1966 headline in the *New York Times*. The following year a Westinghouse executive predicted that there would be a hundred thousand electrics on America's roads within a decade. Even some managers at the major car companies believed that concern over smog would eventually force government to ban the gasoline engine, a belief that led to a renewed interest in electric vehicles in Detroit.

With all the stars seemingly aligned, futurists of the era believed the arrival of the mass-produced electric car was inevitable. Writing in 1969, author Stuart Chase sketched out this placid vision of the year 2000: "[An observer] hears no . . . grinding trucks, roaring motorcycles, screeching station wagons or grunting bulldozers. Along the clear, almost transparent road, faintly luminous at night, comes a fuel cell car, small, quiet, easy to park, shockproof, fumeless . . . an electric truck follows."

Then came the 1970s, the decade bracketed by two major disruptions in America's oil supply. As punishment for supporting Israel in the Yom Kippur War, the Arab members of the Organization of the Petroleum Exporting Countries (OPEC) declared an oil embargo against the United States in 1973. The embargo almost brought America, heavily reliant on imported oil, to its knees. In a country that had always had as much fuel as it wanted, service stations ran out of gas. Motorists in their big cars spent hours waiting in line for a few precious drops; in some areas police were called out to protect tanker trucks from robbers. In 1979, just as the country was recovering from that first oil shock, the new Islamic fundamentalist government of Iran briefly turned off its oil tap, too, sparking a second fuel crisis.

With oil scarce and smog smothering major cities, electric-car enthusiasts were sure that this time America would get the message. In 1975, as owners of Buicks and Cadillacs saw prices at the pump skyrocket, service stations that allowed drivers of electric cars to exchange depleted batteries for charged ones opened in Boston. The infrastructure that would make electric cars as convenient as gasoline-powered machines was falling into place.

At the same time, predictions by experts and policy goals set by frantic officials led Americans to believe that oil was going to be playing a dramatically reduced role in the future. Agencies such as the U.S. Forest Service predicted in 1974 that pollution-causing vehicles would be banned from cities by 1990. Four years later, James Coomer, a professor at the University of Houston, said that eight-cylinder cars would be phased out within a decade. In 1979, President Jimmy Carter proclaimed that by 1990 America would reduce its consumption of oil by 4.5 million barrels a day.

In fact just the opposite happened. As this century dawned, millions of cars were burning a lot more gasoline than in the 1970s, and in America more petroleum was coming from overseas than ever before. Hybrid electric vehicles, machines powered by fuel cells, and other types of cars and trucks had arrived, but still made up only a fraction of all cars sold. Three little words largely explain why we're even more addicted to oil now than when we were in Carter's day: *supply and demand*.

As fuel prices rose in the 1970s, millions of Americans discovered that they could be just as happy with a well-made four-cylinder car as they could driving one with eight. Baby boomers in particular, many of whom were too young then to afford Detroit's heavy, expensive models anyway, turned to fuel-efficient imports from Germany, France, and, most of all, Japan, which at that point was just starting to take on the American market. (Detroit was largely blindsided by the Japanese. In 1970, Lee Iacocca, then an executive with Ford, advised race car driver Jim Shelby to pass on a chance to buy a Toyota dealership. "We're going to kick their asses into the Pacific Ocean," Iacocca said of Toyota. Shelby later estimated that his decision to heed Iacocca's advice cost him about $10 million.)

Smaller cars burned less gas, eventually reducing America's need for foreign oil. At the same time, campaigns to get Americans to conserve energy at home were also working; when millions of people kept their lights off for a few extra minutes a day, the cumulative effect was huge. And while the government never instituted some of the more radical solutions that were being considered—such as the widespread rationing of gasoline—it did establish fuel economy standards that, at least for a while, put a lid on increases in consumption. (Mileage, however, never achieved anything like 400 miles per gallon, the level General Motors executive Alfred P. Sloan predicted in the 1930s.) Cars had even become cleaner,

too, thanks to federal air quality laws that made catalytic con-
verters and other emissions controls standard equipment.

By the mid-1980s, oil prices had fallen closer to pre-oil-
shock levels in keeping with the drop in demand. With the
economic pressure off, interest in developing viable electric
and other alternative cars evaporated. Instead backsliding was
the norm. Detroit, exploiting a loophole in the federal fuel-
efficiency laws, started building gargantuan SUVs that got as
little mileage to the gallon as any pre-shock gas guzzler. By
the late 1990s, fuel conservation came to be regarded like the
Pet Rock, a '70s curiosity long since relegated to the nation's
mental attic. Far from Carter's goal of reducing fuel consumption, petroleum use actually rose 20 percent between 1982
and 1997, and by 2000 half of the oil America burned daily
came from overseas.

Today, however, everything old is new again. Once more
automotive emissions (which lead to global warming) and
rising fuel costs have made alternatives look increasingly attractive. Nearly every major car manufacturer now offers gas-
electric hybrids that burn less than half the gas of comparable,
traditional cars—and those cars are selling. Still, it's not clear
what the future holds for the electric or hybrid automobile.

After all, we have been here before.

Prediction: Your Car Does the Driving

First came the cars. Then came the paved roads. Then came
the traffic jams. By the 1930s—just a few decades after some
had dismissed the automobile as a fad for the wealthy few—
engineers realized that congestion on America's roadways
was soaring. They also realized that the problem was com-
pounded by the fact that the average person was not an excep-

tionally good driver (no matter what the average guy might think). Bad driving caused accidents that backed up traffic and, more importantly, claimed lives. The problem was the motorist. The solution? Let someone else—or, more precisely, something else—do the driving.

The 1939 World's Fair, held in Flushing Meadows, New York, featured the "Futurama" exhibit, which was sponsored by General Motors. A breathtakingly intricate scale model of a city of the "future" (1960), the display depicted an urban core of skyscrapers soaring above cloverleafs and multilane expressways. This vision was surprisingly on the mark save for one small detail: the cars would be radio-controlled from a central location to ensure the smooth, steady flow of traffic.

Instead of thousands of drivers making tens of thousands of imperfect decisions, a traffic controller would keep cars the ideal distance apart, and at the ideal speed, to prevent bottlenecks. Slowpokes at the head of the pack would be speeded up, while crazed lane changers would be tamed—all in the name of getting everybody from point A to point B as fast as possible.

Work on the self-piloted car, along with a lot of other things, was interrupted by World War II but resumed a few years after it concluded. In the early 1950s, the Chicago Transit Authority planned to introduce "automatically guided bus trains" within ten years. Around this time, General Motors built a concept car, the Firebird (not to be confused with the mass-production car that came a decade later), that featured an "Autoglide." This consisted of a mechanical system that would operate the brakes, accelerator, and steering via radio commands transmitted through magnetic strips buried in stretches of "automated" highway. When the driver was on the nonautomated sections, he or she would operate the car the normal way. In 1967, according to one report, "automatic control of cars" was considered "a good possibility for introduction before the end of the century." Experts polled in 1973 predicted that "automatically controlled cars should be in the breakthrough stage by 1980 and should be in large scale use" by 2000.

GM's concept Firebird, being self-piloted and computer-controlled, was also intended to go fast—very fast. In addition to the Autoglide, the Firebird of the '50s and '60s also had a radical new propulsion system based on the jet engine (more on that in a minute). Harley Earl, the legendary GM designer, helped create this "car of the future" and drew up plans, in 1956, for the "Motorama," the "highway of tomorrow" where sleek, low-slung vehicles such as the Firebird would zip along at speeds of more than 100 miles per hour on elevated roadways kinked with hairpin turns snaking around hills and mountains. Thousands of cars would be able to do so without wiping out thanks to high-powered central computers that would communicate via radio with the electronic brains installed in the cars themselves.

The appeal of the automatic-highway concept only grew as America's streets became even more congested during the last quarter of the 20th century. And, in addition to eliminating the migraine-inducing commute, automated driving also promised to increase highway safety—experts say about 90 percent of all traffic accidents can be blamed at least partly on human error—as well as being an economic boon.

Simultaneously, technological advances—such as cruise control—also bode well for the driverless future. Cellular technology could, in theory, help cars communicate with one another, which is crucial in avoiding collisions. Ever smaller and more powerful electronics would make creating the infrastructure necessary for automatic roadways a lot easier than it would have been back in the Futurama days of the 1930s, when electronics were bulkier and less advanced.

But the day when millions of motorists could take their hands off the steering wheel still appears to be at least another few decades off. A lack of political will is probably the main culprit. While government agencies such as the U.S. Department of Transportation (DOT) have supported research into so-called "AHS" (automated highway systems), Washington has been more inclined to stick to the tried-and-true.

In August 1997, for example, the DOT set up an automated highway in Southern California as a demonstration project. For three days, specially modified Hondas and Buicks traveled along a stretch of Interstate 15 in San Diego while a computerized traffic management system, via remote control, handled the steering, braking, and acceleration under real-world conditions. Segments of highway that could normally handle no more than two thousand cars an hour were seeing twice that number during the demonstration. The project was widely hailed as a success—so much so that some headlines even proclaimed that the automatic car had, after sixty years, finally arrived. The following year, the government canceled all funding into AHS research, citing budget pressures.

Retrofitting highways for automatic driving would be incredibly expensive and draw money away from expanding and maintaining the highway system America already has. The public wants its local bit of interstate fixed now, and probably would not have the patience to wait for innovative AHS systems—even if they are ultimately the best chance for freeing up traffic and saving lives. Still, work continues on the self-driving automobile: some carmakers have begun to introduce advanced collision avoidance technology in their luxury models, and many vehicles are sold with computer navigation systems.

And in another huge step, the luxury brand Lexus recently introduced a car that can park itself. Should the driverless future ever arrive, it will probably have come in many small, evolutionary steps, not a big revolutionary one.

Prediction: Cars Will Be Jet-Powered

The rock stars of the early 1950s were America's jet pilots. They went faster than anyone else, literally defied death, and, most importantly, flew some of the sexiest-looking machines ever built. It was almost inevitable that elements of jet aircraft design would be seen in the cars of the day, to appeal mostly to guys who wanted to emulate the swagger of the fighter jock but who didn't exactly have the right stuff. Tail fins back then were practically standard equipment on your typical Chevy.

General Motors' concept car, the Firebird (discussed in the previous section), took things a step further. The 'Bird, which went through several incarnations during the '50s and early '60s (the Firebird I, II, etc.), not only resembled a jet plane, it actually had what was basically a modified jet engine. By now, a lot of people back then thought, all of us would be driving these jet cars.

By the time the Firebird (and a similar British car, the Rover) was built, jet engines had been in existence for about twenty years. The concept is deceptively simple: using turbines, oxygen is sucked into a fuel chamber and ignited, hot gases are expelled out the back, and the plane is pushed forward. In the automobiles, the exhaust was channeled toward flywheels connected to the tires, turning them. With no pistons or the other accoutrements required by an internal combustion engine, gas turbines had far fewer moving parts (simplicity of maintenance was one of the Firebird's big attractions).

The Firebird prototypes were low-slung, bullet-shaped, and sometimes had a huge single tail like a shark's fin in the back. More often than not, the driver was seated beneath an aircraft-style plastic canopy. They sometimes had stubby wing stabilizers protruding from the side, intake vents for the air that helped ignite the car's fuel, and a funnel-shaped exhaust in the rear. Even the steering column was more like an F-18's than a Ford's: in lieu of a wheel, there was a joystick. Most versions of the Firebird looked something like the Batmobile (from the campy 1960s TV show, not the more recent movies).

The day when all of us would be driving jet cars seemed to come a lot closer in June 1961, when one of the men who helped create the Firebird, self-taught engineer Emmet Conklin, took his prototype for a test drive on the streets of Detroit (accompanied by a police escort). Astonished onlookers heard a "high-pitched whine, almost identical to that of a jet aircraft," according to a reporter who went along for what must have been a very thrilling ride, as Conklin drove the Firebird through the city's downtown. "The day when cars like the Firebird become commonplace is not as far away as you might think," Conklin assured his passenger as their ride to the future came to an end.

While not everyone agreed with Conklin's sunny assessment, many experts did. The consensus opinion among Detroit engineers polled in 1961 was that the widespread use of cars without pistons was just five or ten years away. Noting that cars like the Firebird required less space for the engine, supposedly emitted less pollution, and delivered more power than conventional motors, Michael Ference, Ford's vice president for scientific research, predicted in 1967 that there "would be a lot of gas turbines around" by 1985.

Aside from that amble on a sunny Detroit morning so many years ago, however, jet cars were almost never seen again outside of proving grounds and test tracks. Eventually even the prototypes faded away, and all that remains of Conklin's grand dream are a few Firebirds lovingly preserved by aficionados and automotive museums.

The jet car's chief drawback was its price tag: Conklin's prototype cost about $1 million back in 1961, very hefty even for a concept vehicle. Using the exotic materials and engineering the parts that can withstand the heat of a turbine engine makes economic sense when building a multimillion-dollar airplane, but not so much for a consumer product like an auto. Some versions of the Firebird were made mostly of titanium, and the electronic computers and controls used to steer the thing were also very pricey. Jets also gulp fuel, and the Firebird was no exception—the "car of the future" would have required a lot of trips to the gas station of the future.

Prediction: Up, Up, and Away in My Dirigible Balloon

Birds do it, bees do it, and, eventually, people were able to do it, too. The mistake people made for thousands of years was believing that they could do it in the same way as the birds and the bees.

Flying. From the legend of Icarus to the fancies of Leonardo da Vinci, many smart people labored under the misperception that the best way to take to the skies was to mimic hawks and hummingbirds: slap on a pair of knitted wings and start flapping. But what worked so well for birds was never going to result in anything but ridicule—or worse—for any person who tried to emulate our feathered friends, whose bodies, relative to the thrust their wings can produce, are much lighter than a human being's.

Early on, humans figured out how to build craft that could sail on water. The big trick was to make something strong enough to support a person's weight but not so heavy that it would sink. The quality the craft has to possess is buoyancy—meaning that the boat's weight is less than the weight of the water it displaces. A small rock weighing a few ounces sinks to the bottom because it weighs more than an equivalent volume of water, whereas a thousand-ton battleship stays afloat because it is lighter than the water needed to fill the space it occupies.

It took people a lot longer to realize that roughly the same principles apply when considering the ocean of air above us. Certain gases—such as helium—weigh less than the mix of oxygen and nitrogen that makes up the atmosphere, so a closed container filled with such a gas will float. Likewise, warm air is less dense than cold air, so a container filled with hot gases will lift into the sky as well.

That's the secret discovered by the Montgolfier brothers of France, who, in the late 18th century, built a hot-air balloon and launched the era of controlled human flight. For the next hundred years, "aeronauts" bobbed gently along in their hot-air balloons but only had limited control over where the winds would take them. The next big challenge to overcome was building an airship that could not only float but fly.

The arrival of the first internal combustion engines seemed to point the way. Instead of merely being pushed along by the prevailing winds, a balloon outfitted with a motorized propeller could actually be pointed in a particular direction. For years various experimenters fiddled with what came to be known as the dirigible—a balloon reinforced with metal supports and carrying a gondola—until Brazilian Alberto Santos-Dumont flew the first one over Paris in 1898. Santos-Dumont's flight was regarded as a kind of late-19th-century moon shot, a feat of daring and technology that captured the imaginations of millions.

Still, the dirigible did have its detractors. Sir Hiram Maxim, inventor of the machine gun, said in 1903 that the vehicle "would always be at the mercy of a wind no greater than that which prevails at least 300 days in the year," just fourteen years before a dirigible would travel five thousand miles nonstop.

But for the most part, military planners immediately saw the mighty airship's potential, and if anything were too enthusiastic about it. Rudolf Martin, a German writer, predicted in 1907 that dirigibles would soon transport entire armies of half a million men to the battlefield and that the vehicles would forever be superior to the airplane.

The dirigible was also expected to have a profound influence in the civilian sphere as well. Asked to comment on what he believed would be the state of transportation a hundred years hence, Kansas politician John Ingalls in 1893 predicted that "it will be as common for the citizen to call for his dirigible balloon as it now is for him to call for his buggy and boots." In 1890, John Wells, a judge from New York, envisioned the year 2000 as a time when "airships and balloons, dirigible and easily controlled, are a popular means of locomotion." "To consider this present case," read a July 26, 1897, editorial in the *New York Times*, "is to see how enormously the conquest of the globe by civilization would be hastened by the production of an airship. If necessity be the mother of invention, she might even now be 'yearning at the birth' of the dirigible balloon."

Martin, Wells, Ingalls, and many others were commenting at a time when progress in dirigible flight appeared to be proceeding at a pace far greater than that enjoyed by the airplane. Following the first successful, controlled voyage of the machine in 1898, advances were made rapidly, particularly by the Germans. Within a decade, Ferdinand von Zeppelin had built a craft more than four hundred feet long, and the next year the old count helped found the world's first passenger air transport service.

Later, during World War I, "zeppelins" were routinely achieving altitudes of twenty-five thousand feet and carrying substantial loads of bombs. A few years after that, dirigibles began making transatlantic flights with crews of up to two or three dozen people, and by the late 1920s passenger service between Europe and the Americas began. A number of popular magazines of the day had covers illustrated with "cities of the future" depicting stately, cigar-shaped airships gliding past majestic skyscrapers. Even well into the 1930s—after the airplane had already established itself as a reliable, speedy means of freight and passenger transport—dirigible flight had not yet been entirely written off. The Empire State Building even contained a dirigible dock and terminal near its spire.

"Time and the engineers will tell whether the dirigible really has a place in our transportation scheme. Expensive wholesale experimentation is probably just around the corner," C. C. Furnas, a professor of chemical engineering, wrote in 1936. The following year, however, "wholesale experimentation" came to a halt when the *Hindenburg*, filled with explosive hydrogen gas, blew up on approach at Lakehurst, New Jersey. The United States and other nations that had been constructing dirigible fleets all but abandoned them almost overnight.

Tragic though the consequences were—one-third of the approximately ninety-five people on board were killed—the dirigible's advocates argued that the *Hindenburg* crash was more public relations disaster than an accurate reflection of how safe that form of travel had been. The *Hindenburg*, as many know, would have been filled with helium—which is not flammable—if the United States, the chief manufacturer of the gas, had not banned its sale to Nazi Germany. With that hazard removed, the ship might very well have flown on for years; its sister ship, the *Graf Zeppelin*, had, after all, logged more than a million incident-free miles before she was grounded following the *Hindenburg* crash.

Decades after the *Hindenburg*'s fiery death, a few dreamers still believed that the dirigible could fill a niche in transportation. Francis Morse, a Boston University engineering professor, proposed a craft in the 1960s that could carry four hundred passengers in well-apportioned staterooms, a cruise ship of the sky that would be nuclear-powered. Decades later, in the midst of the oil shocks of the 1970s, some proponents saw the dirigible as an energy-efficient and more convenient alternative to freight transport by truck, train, ship, or airplane. Writing in 1979, Adam Starchild and James Holahan envisioned craft that would be up to a third of a mile in length and capable of hauling million-pound payloads. "Given the potential of airships as transportation vehicles, it would not be surprising to see the solutions before the end of the present century," the authors noted optimistically.

Today, some have proposed building huge, wing-shaped lighter-than-air craft that could keep tons of cargo aloft for days on end. Some UFO debunkers even believe that the U.S. military has in fact built such dirigibles, stretching half a mile in length, and that at least a few UFO sightings in the last few years could be chalked up to test flights of these craft. But if the armed forces do in fact have their hands on something like this, so far they've kept it to themselves.

Prediction: An Airplane Won't Take Off

The airplane and the car have a lot in common. Both were perfected around the same time, both revolutionized the business of getting around—and both initially got the Rodney Dangerfield treatment. Just as many experts were sure there was no real future for the car, there were a lot of people who believed that the airplane would never amount to much, either.

This was partly because the principles of airplane flight were a lot more difficult to grasp than the principles behind dirigible flight, which, after all, can be demonstrated using a toy balloon. Airplanes, on the other hand, get and stay aloft thanks to the shape of their wing and the power of their motors relative to their overall weight. But back around 1900, very few people understood both aerodynamics and the internal combustion engine. Instead the skeptics—including some of the most brilliant people of the era—saw planes as nothing more than flimsy, ugly wood-and-canvas death traps.

Among those in the chorus of naysayers was Lord William Thomson Kelvin, the physicist renowned for his work on thermodynamics and after whom the Kelvin temperature scale is named. In 1896, Kelvin said, "I have not the smallest molecule of faith in aerial navigation other than ballooning. . . . I would not care to be a member of the Aeronautical Society." Another prominent skeptic was Thomas Edison, who in 1895 declared that airplane research was a dead end whose possibilities have been "exhausted." Even Orville and Wilbur Wright's father, a preacher, supposedly scoffed at the possibility during a sermon before an Ohio congregation.

But perhaps the most spectacular miss of all was recorded in the pages of the *New York Times* on October 9, 1903, when an editorial appeared noting that "if it requires, say, a thousand years to fit for easy flight a bird which started with rudimentary wings, or ten thousand for one which started with no wings at all . . . it might be assumed that the flying machine which will really fly might be evolved by the combined and continuous efforts of mathematicians and mechanicans in from one million to ten million years. . . . No doubt [aircraft flight] has attractions for those it interests, but to the ordinary man it would seem as if effort might be employed more profitably."

That guess was off by almost exactly "one million to ten millions years": two months after it was made, the world's first working airplane, invented by the Wright brothers, ascended from the dunes of Kitty Hawk, North Carolina.

Still, it would be some time yet before the airplane got the respect it deserved. For years after its invention, heavier-than-air flight was believed to have no future other than as a hobby for those with lots of money and a taste for white-knuckle thrills. Several experts—including some who proclaimed a few years earlier that a heavier-than-air craft would never work at all—were now saying that airplanes would never be practical, powerful, or very fast.

"The popular mind often pictures gigantic flying machines speeding across the Atlantic and carrying innumerable passengers in a way analogous to our modern steamships," the astronomer William Pickering wrote a few years after the Wright brothers' historic flight. "It seems safe to say that such ideas must be wholly visionary, and even if a machine could get across the expense would be prohibitive to any but the capitalist who could own his own yacht. . . . [I]t is clear with our present devices there is no hope of competing for racing speed with either our locomotives or our automobiles."

Nevil Shute, the author of the Cold War classic *On the Beach* and a famous designer of dirigibles, wrote in 1929 that the maximum speed that aircraft would achieve by 1980 would be about 130 miles an hour and its best range would be six hundred miles. (Well before 1980, millions of people had flown on jets that cruise at 500 miles an hour and could travel the length of the United States without refueling. Jet fighters had been built that could travel at multiples of the speed of sound.)

Furnas, the Yale University professor, pointed out in 1936 that the typical airplane's ascent to a cruising altitude of approximately five miles takes "about an hour," so for everything other than transcontinental flights, airliners would fly at a height of just a few thousand feet in order to cut down on travel time. Writing just before the dawn of the jet age, Furnas failed to foresee the arrival of airliners that could soar tens of thousands of feet in just a few minutes.

Flight was believed to have other inherent constraints as well. As the sight of fragile biplanes buzzing rooftops became more common during the early 1900s, landowners began to fret about the possibility of the vehicles crashing into their homes and to ask whether the pilots should be charged a fee for the right to cross "their" airspace. "It is proposed, for one thing, that the [United] States establish 'sky highways' by condemning strips of . . . air, setting them aside and taking title, precisely as though they were land," Philip Ambrose wrote in the December 12, 1909, edition of *Harper's Weekly*. Ambrose went on to quote a New York Supreme Court judge, James Gerard, who suggested that individual property owners be paid by fliers who cross their backyards, and that private citizens be allowed to prohibit airships from traversing their land.

Given all these perceived limitations, it's no surprise that few believed the airplane would ever be useful in military combat beyond perhaps filling a reconnaissance role. Chief among the skeptics was Ferdinand Foch, who led the Allied forces during World War I and whose native France was, ironically, at the forefront of aircraft innovation at the time (equally ironic was the fact that the French government would later name an aircraft carrier after Foch). In 1911, the celebrated commander said that the plane was a "toy" with "no military value."

How quickly was Foch proved wrong. In the ensuing decades, planes became faster and faster. Engines became more powerful and fuselages both lighter and stronger, so more freight and more passengers could be transported. Then came the invention of the jet in the 1930s, and airplanes became faster still, until test pilot Chuck Yeager broke the sound barrier in 1947. That set off an even more vigorous competition among American test pilots throughout the 1950s.

Around the middle of that decade, the first commercial jet airliners appeared, traveling just under the speed of sound—fast enough to get hundreds of passengers from New York to California in five hours. Soon airplane makers began setting their sights even higher.

Prediction: Planes Would Span the Globe in a Few Hours

From the 1930s to the 1960s, the top speed for a fighter plane increased by a factor of about seven. Commercial airliners roughly doubled in speed during the same era. Forecasters, including experts at the RAND Corporation, looking ahead to today figured that the trend would continue, that by now it would take about as long to fly from New York to London as it would to check your bags for the same flight. Since then, airliners have gotten a lot better in many ways but, interestingly, they're not all that much faster than they were back in the '60s. In fact, with more planes in the air, more delays, more invasive security checks, it probably takes us even longer to get there and back than it did our parents and grandparents (and they got to keep their shoes on).

A generation ago, though, making airline travel even faster was considered primarily a matter of making a faster plane. Engineers in a number of countries were fairly confident that the technological challenge, at least, could be handled. Some designed airliners with adapted military engines. Other designers drew up plans for craft that would follow an arc sixty miles up (five times the altitude of conventional airliners) and come drifting back down following a path that would deposit the craft several thousand miles from it starting point. A 1960s RAND study weighed the merits of building jets with nuclear propulsion systems, and nuclear jets were considered to be a good possibility for future flight.

The first step, though, was to build a passenger airliner that could break the sound barrier, an extremely complex engineering problem that almost no private companies wished to take on. (Airliners today travel almost, but not quite at, the speed of sound.) Instead three of the world's biggest governments decided to give it a go.

The United States began drawing up its plans for a supersonic passenger jet around the same time that an Anglo-French joint venture began work on the craft that would eventually become the Concorde. The American project quickly ran into roadblocks, however. The first involved sonic booms. The public found out that since the plane would be flying at greater than the speed of sound, tremendous air pressure would build up on the leading edges of the wings, pressure that was only relieved when the craft punched through that invisible wall. The result would be a burst of sound and energy that would severely disturb the peace of anyone unfortunate enough to find themselves in the plane's flight path. Within a few years of proposing it, the government agreed to restrict the craft to overseas routes—in a stroke, severely limiting the plane's profit potential. Meanwhile, America was trying to land a man on the moon, fighting a costly war in Vietnam, and spending billions on domestic programs. With so many other priorities, Congress eventually dropped the project.

But for the British and French, the supersonic airliner was their moon shot, a chance to strut their nations' technological stuff on a world stage. And when it took off on its maiden flight in 1976, achieving a top speed that was twice a conventional airliner's, the world was impressed. Many believed that the Concorde had set a new standard for air travel and that all planes would be supersonic in the 21st century.

But it was not to be. In the early 1970s, when it possessed the best safety record in the world, there were plans to build more than three hundred Concordes—but then came 1973's Arab oil embargo. As with so many other exotic transport options, soaring fuel prices turned the graceful but fuel-guzzling jets into economic albatrosses overnight. Only twenty were ever built, and the British and French governments, far from recouping the cash they had poured into developing the planes, were forced to practically give them away to Air France and British Airways. A few years after a Concorde crashed in 2001, they were mothballed for good.

The Concorde was a cautionary tale for a lot of aircraft companies. Airlines would probably love to offer two-hour service between London and New York for the masses, but they have also seen that there's a real financial risk involved. If another Concorde-style airplane is ever attempted, some government would once again likely have to be the one to build it.

Prediction: Flying Cars

The world of tomorrow has long been defined by the flying car. For generations, makers of science fiction movies let you know you had entered the "future" by depicting average people tooling around in their airborne sedans. You could see flying cars in the Saturday serials of the 1930s and in movies with Bruce Willis and Harrison Ford fifty or sixty years later. When many of us were kids, we figured that a flying car would surely be in the garage by 2010. Needless to say, we've all learned to deal with disappointment.

A man named Paul Moller might just be our salvation. He continues to pursue a dream that first began percolating while he was a boy growing up in Canada: to create a commercially viable flying car. Though various crackpots and tinkerers have tried to do this for nearly a century, Moller is made of different stuff. A man with a knack for machines and an eye for a good business deal, Moller built a carousel for his friends when he was just eleven years old. Later, he skipped college and went straight into a Ph.D. program in engineering, an almost unheard-of feat. He's made several fortunes—including one in California real estate in the '60s and, twenty years later, another for inventing a popular muffler. More recently, he racked up yet another pile with a small engine he developed.

But everything in his life has been a sideline next to his baby, the Moller Sky Car, a sedan-sized machine that heads aloft powered by enclosed propellers and stabilized with the help of onboard computers. As of a few years ago, Moller continued to make incremental technological progress, building a prototype with quieter, more powerful engines. Investors intrigued by his vision and impressed by his determination continued to finance his work. Yet for all that effort, the U.S. government is not even close to ceding the skies to anything but traditional airplanes. And most observers believe that that's not going to change for at least another twenty-five years or so, if ever.

The seeds for a vision of future skies cluttered with small private aircraft were first sown decades before Moller was born, early in the 20th century. "Over cities, the aerial sentry or policeman will be found. A thousand aeroplanes flying to the opera must be kept in line and each allowed to alight upon the roof of the auditorium, in its proper turn," science writer William Kaempfert predicted in 1913. Twenty years later, the concept had found its way into the pages of popular, general interest periodicals. In the March 1938 edition of *Harper's Monthly*, writer Arthur Train, Jr., described a day in the life of "John Doe," an average American of the year 1988, whose family maintained a collection of vehicles on the roof of their home capable of taking off straight into the air.

The helicopter and its variants, it seems, had particular appeal. Not long after workable models were developed in the 1930s, entrepreneurs eagerly predicted a revolution in transportation—perhaps none more eagerly than Harry Bruno, an aviation publicist who, in 1943, said that "automobiles will start to decline as soon as the last shot is fired in World War II. . . . [The] helicopter will all but replace the horseless carriage as the new means of popular transportation. . . . These 'copters' will be so safe and will cost so little to produce that small models will be made for teenage youngsters. These tiny 'copters, when school lets out, will fill the sky as the bicycles of our youth filled the prewar roads."

A few years later a California company attempted to market the "Helipod," which used six rotating wings akin to helicopter blades to achieve speeds of up to 65 miles per hour. "A pod in every yard may make commuting a breeze," read a caption that accompanied an article on the device, which went on to state that initial tests for the craft were planned for the summer of 1962.

Some of the 20th century's most prominent industrialists also worked to create the personal flier. Glenn Curtiss, the famed airplane manufacturer, was involved in the construction of the "aerobile," a combined car and airplane designed to travel on conventional roadways and in the air. None other than Henry Ford was also enthusiastic about car-plane hybrid, telling a reporter in 1940, "Mark my words. A combination airplane and motorcar is coming. You may smile. But it will come." For years after that prediction was made, designers worked on vehicles that consisted of cars hitched to the bottom of airplanes, craft that would enable the driver-pilot to scoot over traffic jams, land, then "convert" the vehicle back into a car. The Ford company's effort to build a car-plane ended tragically in 1973, following the fatal crash of the Mizar, a hybrid of a Cessna and a Pinto (the car that eventually became a byword for notorious '70s clunkers).

Given the obstacles involved, the dream of taking to the air as a substitute for the highway has proved to be a remarkably enduring one. Using rhetoric virtually echoing that of Bruno and Ford decades earlier, Faith Popcorn, a marketing expert, wrote in 1991: "Imagine the possibilities when the 'personal plane in every driveway' becomes a reality. Under development since 1956, this new individual commuter flier will soon be available, and the time has never been righter [sic]."

In retrospect, it seems obvious that planes or helicopters would have had a tough time shouldering Toyotas aside. Conventional aircraft are notoriously difficult to master, with many pilot trainees having to take hundreds of hours of lessons to earn a license (compared to the weeks of instruction your typical teenager needs in order to learn how to drive). Whatever else its other faults, the car can also travel in nearly all kinds of weather, whereas personal helicopters and planes—as anyone who has been stuck at the airport waiting for a thunderstorm to clear out of Chicago knows—would frequently have to be grounded for safety reasons.

Air traffic control, already a challenge with just a few dozen airliners crossing near an airport at once, would be exponentially more difficult if thousands of sky cars were buzzing around (although powerful, modern computers may be useful in preventing airborne fender benders). And because the consequences of mechanical failure a thousand feet up are far more serious than they are for those tooling down the freeway, aircraft must be built to much higher tolerances, making them a lot pricier than an equivalent land-bound vehicle. (The cost of a personal flier, as well as safety concerns, helped doom Ford's earlier efforts to market the flying car.) Determined though he may be, Moller has his work cut out for him.

Prediction: Hovercraft Will Be Used like a Car

Imagine if you could have the car without the road. Back in the 1950s, some believed that the vehicle of the future wouldn't fly, it wouldn't roll—it would float. By now, according to a few predictions, we would all be gliding along in our personal hovercrafts, levitating above highways that had been given over to flowers and grass. Hovercraft can go anywhere a conventional car can and then some—even over water or a rocky mountain trail. Some even thought that the advent of hovering cars and ships would affect where millions of us would live and work; without the need for paved roads, you could easily commute to work from your home in the woods.

Hovercraft—also known as "ground effect machines," or GEMs—rest on a cushion of air trapped between the bottom of the craft and the surface; they're usually pushed forward and back by a propeller. The bigger the craft, the greater the

height at which it can float, so for GEMs that are big enough
even a landscape of molten lava is theoretically passable. Although the concept was first described in the 1700s, practical,
working models were not built until the late 1950s. Unlike
other novel modes of travel, the hovercraft's journey from
prototype to commercially viable ship was remarkably short.
By the early 1960s, waterborne hovercraft were transporting
passengers in Great Britain (where much of the early research
work was done), and since then, millions of people, especially
in Europe, have traveled on hover-ferries.

Its rapid success fueled the subsequent hype about the hovercraft's potential as a replacement for the four-wheeled auto. A
few years before regular hovercraft service began in Britain, the
Ford Motor Company rolled out the "Levacar," a 450-pound
(driver included) eight-foot compact that glided on a cushion of
air and could hit speeds of up to 100 miles per hour.

The prospect of the car without wheels stretched many
imaginations, including some of the best, like Arthur C.
Clarke's. The science fiction writer-cum-futurist predicted
in the 1960s that the era of the rolling car was rapidly drawing to a close. "[T]here will be a very difficult transition period before the characteristic road sign of the 1990s becomes
universal: NO WHEELED VEHICLES ON THIS HIGHWAY." Clarke
also envisioned a world where, thanks to hovercraft that could
easily sail past the coasts and directly into the heartland,
Oklahoma City could become a major port, and landlocked
Switzerland a shipbuilding nation. "All this is very bad news
for San Francisco, London, New Orleans . . . or any other port
city you care to name," Clarke wrote.

In 1970, marine scientist Lloyd Stover said that commuter
hovercraft might soon be plying the waters off cities such as

Washington, D.C., and that by 1985, "large oceangoing [hovercraft] may be the major means of transportation along the surface of the world ocean." That same year, air-cushioned "personal capsules" were envisioned ferrying people around major metropolitan areas within a few years. And the Soviet Union spent decades developing what came to be known as the "Caspian Sea Monster," a five-hundred-ton GEM that was to be used as a missile platform and troop transport. The Russians abandoned the effort after the titanic machine crashed in 1980.

While hovercraft are used today for a wide range of applications—as ferries, in the military, and for certain specialized industrial tasks—there is little chance that they will replace conventional cargo ships, much less automobiles. Hovercraft suffer from a limitation that has doomed many other promising transportation technologies: high fuel costs. To achieve even moderate speeds hovercraft require a lot of horsepower, and the dramatic rise in gasoline prices during the latter part of the century largely explains why more GEMs were not built. Conventional GEMs, powered by what are basically airplane engines, are also incredibly loud (imagine being awoken in the morning to the sound of the neighbor's Leva Car warming up!) and kick up a lot of debris when operated over land. Today you may be able to hitch a ride on a hovering craft, but it's doubtful you'll ever use one to get to the supermarket.

Prediction: We'll Get on the Bus—or the Intercontinental Subway

Mass transit makes a lot of sense—it's better for the environment, cheaper per capita than maintaining a car, and spares the commuter the hassle of sitting in traffic. A generation ago planners figured that we Americans here in the advanced, cutting-edge 21st century could easily be coaxed from behind the steering wheel and wedge ourselves into a seat on the high-tech and convenient trains, trolleys, and buses of the future. As it turned out, though, for most modern Americans taking mass transit is kind of like eating your veggies—something we should be doing more of but just don't. We're even more hooked on our cars now than we were forty or fifty years ago.

That's not where we started. At one time the only mechanized form of transportation that existed was intended for the masses, and public transport was supposed to be the successor to the horse and buggy. From the late 19th century up until World War II, many Americans who wanted to travel between and within cities relied on the train and its cousins, the trolley and the subway; only a relatively small number had their own car. Public transport systems, it was believed, were to dominate the scene far into the future. "[S]urface travel will be an oddity in New York in 20 years," John B. McDonald, who built Manhattan's first subway, predicted in 1903. "Unless all signs fail, another decade will see America gridironed with trolley lines from ocean to ocean," writer Alexander Hume Ford proclaimed in the May 29, 1909, edition of *Harper's Weekly*.

But as the country suburbanized, local governments began spending more and more for roads and less for buses and trains. Mass transit and suburbia didn't mix: if you could afford a home with a yard, you could afford a couple of cars, too. Why bother with a bus pass when you had a Buick?

Still, some experts believed the country was approaching a tipping point, that people were getting sick of the American car culture and only stuck with it because there was no alternative. Build mass transit suited for low-density suburbia, some thought, and you'll see the seats fill up.

"The so-called infatuation of the American male with his automobile has, I think, probably reached its peak already and may be on the decline—except among the very young and the disadvantaged who have not yet tasted its monoxide pleasures," a physicist, Robert Ayres, wrote in 1968. "I doubt that the average suburbanite feels an automobile is tantamount to personal freedom; it has become a chore, like other things."

Ayres championed various alternatives to augment the congested freeways, including sky "buses," automated electric taxis, and moving sidewalks in downtown areas—ideas that were all being seriously entertained at the time.

In 1968, President Lyndon Johnson submitted a report to Congress detailing several of these options. One was known as "personal rapid transit" (or PRT), a category broadly applied to a number of different systems that all featured private, automatic cars of some kind that could be summoned on demand and would move the passenger to any one of numerous stops along a guide rail. An electronic call box system would be used to order up the vehicles, which, Johnson's report said, would move between one thousand and ten thousand people an hour at a speed of about 50 to 75 miles per hour. The system was designed to serve cities and larger suburbs, while a variant, a so-called dual mode system featuring cars that could travel on conventional roadways to the guide rails, would serve less densely populated areas. It was a kind of combined taxi-rail service ideally suited for the burbs.

Getting around within the city, meanwhile, would be accomplished using moving sidewalks, essentially escalators that convey people horizontally at a speed greater than a brisk gait and equal to what a car negotiating downtown rush-hour traffic could manage (these are in use today at many large airports). The Johnson report noted that while the basic technology existed to create moving sidewalks, some hurdles would have to be overcome, such as making sure the slow and the clumsy didn't get their ankles ground into mulch if they didn't step onto the things correctly. Forecasters figured those problems could be solved, though: in 1979, experts looking twenty years ahead predicted that "to some extent automobiles and other forms of vehicular mass transit may be replaced by moving sidewalks."

Other systems were proposed that were closer to traditional mass transit in concept—that is, they were meant to move hundreds or thousands of people at the same time between two points—but revolutionary in execution. Some envisioned trains levitated above the rails using magnets or air cushions and that could move at speeds of 600 miles per hour (far faster than the bullet trains now used in Japan and elsewhere). Another pie-in-the-sky vision was the "Planet Trans," a subway linking New York and Los Angeles that would travel between the two cities in just two or three hours (this plan was almost immediately scrapped when it was determined that the system would only be economically feasible if the equivalent of the entire population of one of those cities used the train every day).

In 1980, the Los Angeles City Council approved "in principle" a $175 million plan to build a 2.9-mile "people mover" in the city's downtown consisting of a series of transit ways mounted on piers and capable of transporting seventy-two thousand commuters a day. A year earlier a similar system was proposed to link shopping malls in Carlsbad and Oceanside, California. And even in Houston—a city that more than almost any other derived its wealth from the gasoline-powered auto culture—officials were considering building a high-speed rail link, even though the city's leadership had been told not to expect federal money to assist with the project.

Yet for all the alternatives being mooted throughout the 1970s, the American driver remained almost as devoted to the conventional automobile at the dawn of the 21st century as he was at the middle of the 20th. While some cities instituted fairly ingenious mass transit schemes, for example, the most radical, such as personal rapid transit and moving sidewalks, never caught on, and only about 5 percent of the American workforce used mass transit of any kind on a regular basis in 2008. And a large percentage of those commuters are concentrated in big cities such as New York and Chicago.

The early 21st century, in short, looked much like the mid-20th century, only more so: more cars burning more gasoline while sitting in even bigger traffic jams, more airline flights leading to longer and more frequent delays, and a continuing neglect of mass transit systems. What would George Jetson think?

Chapter Three

SCARCITY AND OTHER DISASTERS

The year is 2010. Half of the world's nations are embroiled in wars over water rights. Arizona, Utah, Colorado, and large swathes of California have been virtually abandoned due to a lack of water. Governments in the industrialized nations have begun a program of enforced sterilization, allowing only those who can demonstrate their ability to support their offspring the right to reproduce. Chicken, beer, and bread are all luxury items, served only at high-end restaurants catering to the wealthy few. There's a lot less traffic nowadays, but that's only because cars have become prohibitively expensive. Private bedrooms are also a bygone luxury; as in the 19th and early 20th centuries it's once again commonplace for two or three siblings to share a bed. The cities have begun to decay as people scramble for prestigious jobs working farms in the countryside, where there's plenty of food and room.

Something like this scenario was foreseen back in the 1960s by a group of scientists and public intellectuals who predicted doom for the planet unless mankind collectively took action against a widespread, looming threat. While many of these same thinkers were also worried about pollution, nuclear war, and social unrest, one bogeyman was the scariest of them all: the cute, cuddly baby.

Or rather, billions and billions more of them. Population projections from the time indicated that there would be something like 6 billion people on the planet by the early 21st century, or a doubling of the population from the late 1960s level. Meanwhile, other estimates of existing stocks of everything necessary for even a modest quality of life—from crops to copper—showed all of these resources at the beginning of a steep, sharp decline. Many thinkers at the time believed an abrupt reduction in the number of mouths to feed was inevitable—ideally through voluntary means, but most likely through famine and disease. The global baby boom seemed poised to bust in the most horrific way possible.

Another, less pessimistic group also bought into the idea that humanity would soon outstrip the planet's conventional sources of food, water, and arable land, but believed as well that radical new technologies would save the day. If we couldn't grow enough food, these people believed, well, no problem: we'll "manufacture" protein in factories. Fresh water would be piped in from melted icebergs or come directly from the oceans after being treated at nuclear-powered desalinization plants. Some believed, meanwhile, that many of us would develop a taste for plankton—a yucky alternative to grilled chicken, perhaps, but one that at least had the virtue of being plentiful.

Today, tens of millions of people do not in fact have enough food and water. Resources in many corners of the planet have been exploited to the point where they are nearly depleted, and continued high population growth in some parts of the world would indeed be as disastrous as the doomsayers from the '60s onward predicted. But for a variety of reasons, the picture today is not nearly as bleak as those who agitated for "zero population growth" believed it would be a generation ago.

Prediction: Too Many Mouths to Feed

"The battle to feed humanity is over. In the 1970s and 1980s hundreds of millions of people will starve to death in spite of any crash programs embarked upon now."

So went Paul Ehrlich's unqualified declaration of disaster in the 1971 edition of his bestselling book *The Population Bomb*, which helped convince millions of people that the worst thing one could do to the environment was to add another kid to it. Ehrlich, a Stanford University biologist, based his conclusions on some disturbing trends that all appeared to be converging right around the time he wrote his book. One of the most frightening of those trends was sheer exponential growth.

Everything else being equal, the larger the human population is the faster it grows. It took tens of thousands of years for the number of people on the planet to reach a billion, a milestone achieved around 1850. Yet in just the next 150 years the population had increased to six times as many. When Ehrlich wrote *Bomb* in the late 1960s, most projections indicated (accurately, as it turned out) that the population would double in just thirty-five years.

And the fastest growth was occurring in the poorest parts of the world, places where people were already struggling to feed themselves. Erhlich noted that even the rich world was primed for catastrophe unless the birth rate began to equal the death rate—thereby erasing a net increase in the number of people—in such places as India, Nigeria, and Brazil. The millions of "surplus" mouths to feed would be nothing less than apocalyptic. In light of this, modern medical care was not a boon to humanity but actually a scourge, saving millions of infants who would only live long enough to die an agonizing death from malnutrition.

This brings us to the other disturbing trend Ehrlich highlighted: lagging farm production. He wrote that around 1958, "the stork passed the plow," meaning that the rate of increase in food production began to fall behind the rate of increase in people production. "In 1966, while the population of the world increased by some 70 million people, there was no compensatory increase in food production. . . . Only ten countries grew more than they consumed."

Far from being a lone prophet of doom, Ehrlich's book was merely one of the more concise expressions of a widespread belief that humanity would be unable to feed itself for very much longer. Prominent scientists such as Linus Pauling and Jonas Salk, inventor of the polio vaccine (and therefore an indirect contributor to the overpopulation problem), signed a full page newspaper ad in 1968 picturing a baby labeled with the words "Threat to Peace." The year before, in 1967, the administration of President Lyndon Johnson received a report showing that worldwide famine was "inevitable" despite efforts to ramp up food production.

Things were expected to be so bad, in fact, that it would either be illegal or, at least, in poor taste to drink liquor, feed a dog, or smoke cigarettes. Cropland was going to be so scarce that using it for anything else—such as growing tobacco or making kibble—was going to be considered an unconscionable waste, according to one mid-1970s prediction. Around the same time, writer Paul Dickson, in his book *The Future File*, stated quite unequivocally that "it is fact, not fiction, to say that food is going to cost substantially more in the year 2000." A few years later, a poll conducted by *Omni* magazine revealed that most readers believed steak would cost fifty dollars a pound in the America of the early 21st century.

Although fears about global starvation peaked (at least in the public consciousness) in the '60s and '70s, a few stray voices, like Lester Brown's at the Worldwatch Institute, continued to sound the alarm as recently as 1997, when Brown declared that "food scarcity will be the defining issue of the new era now unfolding."

Overpopulation fears had an intuitive appeal to laypeople, who could immediately grasp the problem resulting from ever-increasing numbers of people and stagnant food production. College students across the country formed "zero population growth" (ZPG) clubs in the early 1970s. Books like Ehrlich's detailing the problem sold hundreds of thousands of copies.

It also spurred some people to contemplate radical solutions. Nowadays politicians love your kids (at least that's what they say). Everything they do in Washington or your state capital is seemingly to benefit "working families," and another little mouth to feed is also a nice tax deduction. But back in the 1960s and early '70s, when scholars believed we were about to breed ourselves into oblivion, it was thought that today the government was going to be actively involved in efforts to shoot down the stork—even if it meant taxing you more once Junior came into the world.

Still, some advocates tried. Around 1970, U.S. senator Robert Packwood and Representative Paul McCloskey introduced legislation that would allow tax deductions for no more than two children in a family. A *Life* magazine article published around that time speculated that not only would deductions for "extra" babies be phased out, but special taxes would be imposed on families having a third child and on every child thereafter. "Luxury taxes" would be applied to diapers and formula.

These were hardly the most radical suggestions. Ehrlich, for instance, quoted a scientist who wanted to seed the water supply with birth-control agents (an idea that even Ehrlich dismissed as impractical). He also supported an effort to impose mandatory sterilization on all men in India with at least three children and believed that America should make financial aid to poor countries contingent on the recipient nation's efforts to control its population growth. (This is an interesting contrast with the situation as it exists today, where a bloc consisting mostly of fundamentalist Christians has helped pressure the U.S. government into withholding aid from charitable organizations that promote abortion.)

Forty or fifty years ago some even expected the Catholic Church to relent on its famously uncompromising stance regarding birth control. In the early 1960s, Dr. John Rock, a gynecologist and devout Catholic, drew up a birth-control plan that he said won the support of 95 percent of the Catholic laity he polled who expressed an opinion on it. Other commentators also believed the church would back down in the face of a burgeoning demographic crisis.

Many of the people warning the world about runaway population growth forty years ago pointed to India as an example of what things would be coming to unless extreme measures were taken. In the mid-1960s the subcontinent experienced two massive harvest failures back to back, forcing the country to import huge amounts of food. Citing a 1959 report released by the Indian government, writer Dennis Gabor projected that agricultural production would almost certainly "lag behind" the rate of population increase, resulting in millions of deaths. Picking up on this theme, another writer, Ferdinand Lumberg, believed that worldwide population projections of 6 billion people by 2000 would prove too high because hundreds of millions would have starved to death by then. A 1975 report by the Indian National Committee on Science and Technology predicted that per capita availability of food would drop by a third in the following thirty years, barring a "radical turnaround" in agriculture.

Yet a radical turnaround is just what occurred, not just in India but across the globe, which is why Ehrlich and so many others proved to be so wrong. While readers may be quick to point out that millions have died of starvation since *The Population Bomb* was first published a generation ago and millions of people remain malnourished today, Ehrlich and like-minded thinkers were predicting something much more sweeping than periodic bouts of famine in the Third World. He and others believed that the global agricultural system would be unable to produce enough calories to sustain the more than 6 billion people on the planet today. As it turned out, this is simply not the case.

Experts today have concluded that modern famines and food shortages occur not because of a lack of supply but because of problems of distribution. These problems are often caused by factors that have nothing to do with farmers' ability to coax enough grain from the soil.

For example, in 2008, nations across the globe experienced food riots as the price of basic commodities such as flour nearly doubled from the previous year. This was not because the world's farmers did not grow enough, however, but rather, according to most experts, because a large amount of farmland in the United States was being set aside for corn that was converted into ethanol. Food that would otherwise wind up on plates was instead heading into gas tanks. Other famines, particularly in Africa, have been "created" by corrupt government officials stealing ample supplies of food aid.

Meanwhile, India, far from being a nutritional basket case, saw its wheat yield more than triple in the last decades of the 20th century, a time when its population merely doubled. In fact it has become a net exporter of food, like many other Third World countries that once relied on huge infusions of aid to feed themselves. At the same time, in developing nations such as Vietnam, the percentage of a family's income spent on food has dropped dramatically (even though this country too experienced rapid population growth). Food has indeed become so cheap that the governments of industrialized nations, particularly the United States and nations in Western Europe, are still subsidizing farmers to the tune of billions of dollars a year because the farmers would not make enough money selling their goods on the open market. As Alex McCalla and Cedar Revoredo noted in their 2001 paper, "Prospects for Global Food Security," wheat prices in that year were the lowest they had been in a hundred years. Today many poor countries aren't asking for planeloads of food aid but for the right to sell their less expensive cheese, fruit, and grain to the rich world.

In fact, for probably the first time in history, many societies aren't worried about starvation—they're worried that their citizens are getting too fat. If the population pessimists from the late 1960s could travel in time to present-day America, they'd probably be astounded by the number of young kids waddling out of the grade schools. And obesity is reaching epidemic proportions not just in the United States—where belts had begun straining as far back as the 1950s—but in Europe and even in parts of Latin America and Asia; in 2008, an estimated 25 percent of Chinese were considered overweight. Instead of contending with rickets and other starvation-related diseases, doctors in these places are predicting that hundreds of millions are at risk from heart disease and diabetes, illnesses suffered by people taking in too many calories and not getting enough exercise.

So how exactly did we win the "battle to feed humanity," as Ehrlich described it forty years ago? Much of the credit belongs to an Iowa farmboy who has spent his entire life thinking about ways to get more crops from every acre, particularly in poor countries.

Born in 1914, Norman Borlaug studied plant pathology at the University of Minnesota before joining an effort sponsored by the Rockefeller Foundation to improve farming techniques in Mexico. Within twenty years, by the mid-1960s, Borlaug had helped develop a hardy dwarf variety of wheat that was resistant to pests and diseases, resulting in yields three times higher than had ever been achieved before. Borlaug's innovations, which included the use of modern fertilizer and irrigation techniques, were soon copied across the developing world, an effort that became known as the "Green Revolution." By helping prevent tens of millions of deaths from the starvation predicted by Ehrlich, Borlaug is widely credited with saving more lives than any other person in history.

And even as food production has risen, birth rates in many places have dropped—despite the fact that in most countries, governments did not force people to have fewer babies (China, with its one-child policy, is a notable exception). In fact, in many countries the problem today isn't too many babies, but too *few*. In Western Europe, Japan, South Korea, and other parts of the world, population growth has fallen below the "replacement rate," meaning that those countries are shrinking and, simultaneously, getting older. Even China, with its billion-plus souls, may be facing a crisis by around 2030 or so, when a shrinking workforce will be supporting a much larger contingent of retirees. By 2008, the mayor of one Italian town had even resorted to paying couples a "bonus" of several thousand dollars for every child they brought into the world. Ehrlich and others believed that today there would be too many people and they would be too skinny. As it's turned out, though, we are both too fat and too few.

Prediction: The Food Pill and the Algae Sandwich

At the end of the classic 1973 science fiction film *Soylent Green*, a disaffected cop played by Charlton Heston is seen running through the streets, announcing a grisly discovery: "Soylent Green is peee-pul!"

The movie's title refers to a food substitute used extensively in the early 21st century, by which time the population of the world had exploded. Forty million people lived in New York City and real food was so scarce that the sight of fresh produce could literally make a grown man cry. The government fed the teeming populace Soylent Green, but lied about what all those cardboard-like wafers were really made of: human bodies. That was one rational yet gruesome way to tackle a food crisis—if the problem was too many people and not enough food, then make the former into the latter.

Outside the realm of sci-fi, however, many scientists were in fact doing research into food alternatives that might not have involved using your neighbor as an additive but were still pretty bizarre. One of the most widely suggested of these alternatives came from lakes and oceans, but it wasn't fish—it was fish food, specifically algae and plankton. Strange as it may sound, chowing down on pond scum does have a certain elegant logic in a world that was supposed to be running out of food.

The act of eating is essentially a transfer of calories from the food being consumed to the consumer. If you have a steak, then you are eating cattle that was fed grain that fed off soil nutrients. The meals eaten on each rung of this ascending food chain involve an intake of energy, some of which is lost at each step. What this means is that a lot more people can be fed with the grain destined for the cow than can be fed after that cow has been converted into rib-eyes and New York strips. The closer you are to the bottom of the food chain, the more "energy-efficient" your eating habits.

You therefore could not get much more energy-efficient than the plant matter one finds clinging to the walls of an aquarium tank, which is why an algae and plankton diet seemed like a magic solution to impending mass starvation. As far back as 1953 scientists were already suggesting that we might one day have to turn to this most elemental of foodstuffs. In that year Hans Gaffron, a professor of biochemistry at the University of Chicago, pointed out that an acre of water covered with a special growth medium could produce five or six times more calories from algae than a similar-sized field planted with soya or other types of plant matter. Should mass hunger be looming, Gaffron believed homes might contain rooftop algae beds, which would yield "a dark-green paste with a bland, grassy flavor. Simple cooking make the food's 50 percent protein and 10 percent carbohydrate content available for nourishment." (Hungry yet?)

Around the same time, other scientists envisioned a different solution, with water pumped straight from the sea, treated with some nutrients to stimulate plant growth, then passed through a centrifuge so that the resulting plankton could be collected and pressed into, well, plankton burgers. Physicist and author A. M. Low believed that the first experimental plankton processing plants would be built in California or some other sunny place "in the near future"—that is, around 1960. Still another, even more fanciful method involved the construction of robot "whales" that would constantly troll the oceans, scooping up millions of tons of plankton with their mechanical mouths.

In a world where nutrients would be at a premium, the definition of a "pest" was expected to change radically. In 1966, for instance, an insect expert at the University of California–Irvine, Ronald Taylor, held a press conference during which he fed reporters fried caterpillars, which one journalist described as tasting just like bacon. Like plankton and algae but only slightly higher up the food chain, bugs too are both plentiful and rich in protein. Taylor helpfully listed the more nutritious of the critters for the assembled reporters, insects such as grasshoppers, bees, ants, grubs, termites, and even wood lice (cooked into a sauce).

A decade earlier, author Victor Cohn quoted authorities who believed that meat could be much more plentiful and less expensive, provided you looked beyond the poultry farm or livestock ranch to find it. Depending on what part of the world one hailed from, kangaroos, prairie dogs, camels, and the agouti—a South American rodent—could all one day satisfy that carnivore fix.

When these predictions were made, in the mid-20th century, the chemical industry was producing many of the sexy, headline-grabbing innovations, much like the computer and biotech industries do today. It was during this era that the cliché "better living through chemistry" was born. Advances at the time also promised, incredibly, better eating through chemistry. Why grow or raise your food when you can build it?

As with gaining sustenance from pond scum, getting meals from test tubes was based on some fundamental, commonsense scientific principles. Everything we eat is just differing chemical combinations of very basic elements, such as carbon, nitrogen, and iron. Given that scientists were already successful in manipulating such basic elements to produce a whole range of new products—from plastics to nylon—it didn't seem so farfetched to create tasty, completely artificial alternatives to sea bass or wild rice, either.

As far back as 1923, the scientist J. B. S. Haldane predicted that one day food will simply be "built up" from very simple substances, such as coal and atmospheric nitrogen. "Synthetic food will substitute the flower garden and the dunghill and the slaughterhouse, and make the city at last self-sufficient," he wrote. But Haldane, apparently a cautious type, did not foresee this happening until about the year 2040 or so.

Chemists in later decades, giddy with their success, were a lot more optimistic. Low believed that a chemical process could be devised that would soon make cellulose—the basic component of grass and wood—easily digestible for human beings. An industrial chemist in the mid-1950s, Jacob Rosin, predicted in his book *The Road to Abundance* that synthetic fibers would soon be developed and assembled to create a product identical to red meat. Around that same time, writer Fritz Baade predicted that in the near future, coal and petroleum could be chemically converted into protein and fatlike substances.

Yet another variety of manufactured sustenance was that good old sci-fi standby, the food pill, which was expected to be a staple of the typical diet as far back as the late 19th century. Back then, a French chemist imagined a time when everyone carried around their tablets of nitrogen, spices, fats, and starch, food that could be produced in quantity regardless of drought or blight. Although similar, food pills were not quite the same thing as vitamin tablets, which are merely supplements to a conventional diet. Instead food pills were intended to be the diet. This idea gained a certain cachet with the advent of the U.S. space program in the 1950s, as experts worked to devise compact food sources that could keep astronauts alive on long voyages aboard cramped spacecraft.

Still another sideline in the manufactured food movement involved taking conventional foods and somehow treating them so that they would not spoil, cutting down on waste. Today, of course, the use of chemical preservatives is commonplace, which is why Twinkies and potato chips can stay fresh on the convenience-store shelf for so long. Chemical preservatives are also notoriously unhealthy (nutritionists today urge us to stick to fresh foods as much as possible), but not nearly as troubling to some as one suggested method to keep our lettuce crisp for up to a year: radiation.

Shortly after the first atomic bomb was developed at the end of World War II, some scientists began to see a silver lining in the mushroom cloud. While radiation was already known to be lethal, in judicious doses it could also be a boon to mankind, many experts believed. And radiation is, in fact, sparingly used today as an alternative to pesticides for fruits, vegetables, and spices. A quick jolt of gamma rays effectively sterilizes food, killing off the bacteria that causes spoilage. Produce can be shipped on long trips to faraway markets thanks to this method, which is why someone in New York can enjoy Hawaiian pineapple pretty much year-round.

To this day, the process makes a lot of people more than a little squeamish (some ritzy grocery stores won't even carry irradiated products), and as a result the method is not used as extensively today as it theoretically could be. But back in the 1950s, when scientists were working overtime trying to devise so-called peaceable uses for the atom, they envisioned a future where refrigeration would no longer be necessary and nuked potatoes could stay in the pantry for years without sprouting eyes.

A 1954 *Wall Street Journal* article said that one day, thanks to irradiation, four-month-old bread would still be edible and soldiers would go to the front with cooked steak in their rations. Cohn interviewed a University of Michigan scientist who irradiated chopped meat and placed it in an airtight container, then stored it at room temperature. A year later the meat was still "fresh and juicy." The scientist neglected to say whether it had grown an eye.

Prediction: Water, Water—Anywhere?

In the 1950s experts began looking at demographic trends in the United States and compared them to a map, immediately noticing a crisis in the making. The West and Southwest, where what was once called the Great American Desert is located, were filling up with people who wanted thick green lawns and swimming pools. Meanwhile, in the developing world, populations were also exploding in places with dry climates. Those experts predicted that sooner or later something would have to give.

And in more and more places, that day of reckoning is rapidly approaching. By 2008, the water level at Lake Mead, the primary supply for Las Vegas, was dropping below the point at which pumps could get to it. Once-vast lakes in sub-Saharan Africa and elsewhere had shrunk by half or more. Aquifers—huge, underground natural reservoirs—from Beijing to Nebraska are being sucked dry faster than rains can replenish them.

The situation is dire even in areas that, up until recently, had always had ample supplies. Persistent drought coupled with a huge increase in the population in the Atlanta area has led to shortages so severe that local governments were recently sparring in court over water rights. The snowpack in the mountains of the Pacific Northwest—the primary water source for millions—is threatened by global warming. Even in tropical Florida, famous for its almost daily, monsoonlike rains, a lack of water is starting to hamper economic growth.

Experts decades ago were correct in predicting that something like this would happen. Where they went wrong, however, was in overestimating the wisdom of the politicians responsible for dealing with the problem. Despite the clear warning signs, policies are still in place that allow millions of gallons to be wasted every day. To maintain aquifer levels, for example, rainwater must be absorbed by the soil. Yet we continue to pave over more and more of the country, allowing this source to drain away into the oceans. Politically connected farmers in such places as California and southern Spain use billions of gallons every year but pay a pittance in fees for it. Nor is their produce a vital, or even logical, part of their economies. Farmers contribute only a tiny fraction to California's huge gross domestic product, while some of their counterparts in historically arid southern Spain are raising thirsty, "wet weather" crops, like watermelon.

While today we seem to be applying a lot of ingenuity in wasting water, fifty years ago one scientist devised a clever method to provide for burgeoning Southern California's needs. John Isaacs, a researcher at the Scripps Institution of Oceanography, proposed using icebergs to quench the Golden State's thirst. Borrowing an idea from 19th-century arctic explorers, who used chunks of floating ice to replenish their ships' stores, Isaacs believed we would one day send tugboats to the Antarctic icebergs that periodically drift to the tip of South America, a region close to the powerful Humboldt current.

The tugs would drag the iceberg into the current, release it, then follow the massive island of ice as it made its way north to near the California coast. There the tugs would recapture the berg and tow it back out of the current to a point near the shore. The iceberg would then be fenced in using a special waterproof material and allowed to melt. Since freshwater floats on seawater, the resulting lake could easily be tapped to provide a water source of about 10 billion tons. The plan,

Isaacs reckoned, would cost about $1 million (in 1956 dollars) but would yield about $100 million in water—far more economical than, say, using tanker ships for transport.

The general concept had an enduring appeal. Twenty years later, two U.S. government scientists, W. F. Weeks and W. J. Campbell, devised a plan to construct a "supertug"—a ship that would have about two-thirds the power of an aircraft carrier—to haul Antarctic bergs directly to nations in the Southern Hemisphere, such as Australia (a large percentage of which is a virtual desert). Although they acknowledged that their plan could be costly, those costs would be far outweighed by the benefits: a typical iceberg contains enough water to irrigate six thousand square miles.

Yet another, related solution involved transporting fresh water in huge, five-mile-long plastic bags that would contain ten million tons of liquid. In the late 1960s, a University of California researcher, Roger Revelle, envisioned the supply coming to relatively dry Baja California from the Columbia River via this method by about 1984 or so.

But physically moving supplies from one place to another—whether in iceberg form or in a giant baggie—was not the preferred method for producing more water forty or fifty years ago. Instead most engineers and scientists were sold on the concept of desalinization plants, facilities built near the ocean that would filter out the salts and other minerals from seawater to produce stuff suitable for drinking. Some envisioned the desalinization plants serving other uses, too. The extracted minerals would be sold commercially, while deuterium, a type of hydrogen found in seawater, would be used as the fuel for the nuclear fusion plants that many believed would be up and running by now. (Getting electricity from fusion will be discussed in another chapter.) This being the mid-20th century, engineers also believed that the desalinization plants would be nuclear-powered, resulting in a facility that produced an unlimited supply of water powered by an inexpensive source of energy.

The main reason why the world is not dotted with more desalinization plants today is the cost. The facilities are expensive to operate and do consume a lot of electricity; building dedicated nuclear reactors for the plants proved to be a lot more complicated today than experts believed it would be back then. Desalinization is presently in widespread use in places like Saudi Arabia, but that's only because the dry but oil-rich kingdom can afford it. For most countries, the process simply does not make economic sense.

Another camp, meanwhile, focused not on novel sources of water but on ways to cut down on use. What spurred these thinkers were troubling projections indicating that the United States was on the cusp of a water crisis as severe as the energy shortages of the 1970s. According to a 1975 forecast by the Producers' Council, an international NGO, pollution and misuse were expected to lead to a severe shortage of drinking water in the United States in the 1980s. One expert in the late 1960s believed that unless desalinization became widely used, people would be forced to leave the state of Arizona— and this at a time when the state had only a fraction of the population it does today. A few years later, in 1972, a paper titled "Population and the American Future" predicted that while there would be enough to meet basic needs at the turn of the century, optional uses of water would have been phased out: "lawn watering has gone the way of leaf-burning as an outdoor activity," the paper stated.

The forecasters fifty and sixty years ago who worried about depleting water supplies were definitely on to something, even if their time frame was off. Today a number of economists and military planners believe that unless something is done, the coming decades will see wars fought over water supplies. The two biggest countries in the world, China and India, are facing potentially crippling shortages. California, the world's eighth-largest economy, has been experiencing more and more droughts. We may yet live to see icebergs off the coast of Los Angeles.

Prediction: Not Enough Juice and Other Mineral Deficiencies

The late geophysicist M. King Hubbert enjoys oracular status among many experts who spend their lives thinking about energy. In the late 1940s, when the world seemed awash in oil, Hubbert predicted that the petroleum era would be relatively short-lived. In 1956, Hubbert made another stunning prediction: U.S. oil production would reach its height by about 1970 and then enter a long decline. "Hubbert's peak," as it came to be known, was entirely accurate. Today variations of his statistical models are still used to determine the life span of various oil fields around the world.

A straightforward analysis of the oil market would seem to indicate that, over time, the price of oil would steadily rise, as many predicted in the 1970s, so much so that we would be forced to switch to alternatives. But, as we saw in an earlier chapter, oil prices spiked thanks to an "artificial" shortage in the 1970s, came back down in the last decades of the 20th century, then, as huge countries like China and India started to consume more themselves, began shooting up again.

But to many thinkers decades ago, the vicissitudes of the

oil market were to have been old news by now. Instead nuclear power was supposed to be the source of choice in the industrialized world, generating electricity that, as a sloganeer once famously said, would produce energy "too cheap to meter." In 1969, writer Stuart Chase was one voice among many who described a 21st-century world where safe, efficient nuclear power plants would lead to unparalleled abundance. For twenty years, from the 1950s, "fission" was the answer given whenever anyone mentioned dwindling supplies of fossil fuels.

Environmentalists, however, were much more excited about the prospects of so-called renewable sources, including solar, wind, and geothermal energy—especially solar. As far back as the 1950s, experts predicted that vast arrays of solar panels would be festooning America's desert West, with electricity being sent across the country via a greatly improved electric grid. Wave generators placed in the oceans were expected to be another, bottomless source of kilowatts.

The problem with these technologies has primarily been their cost, at least relative to fossil fuels. To give these new technologies a chance, governments have subsidized them and offered utilities other financial inducements to encourage their use. At the same time, solar and wind technology have vastly improved, becoming less and less expensive every year—even without government subsidies. With the cost of oil rising sharply recently, and with governments pressured to cut down on carbon emissions, alternative energy sources may finally be providing a much larger share of the world's power.

Or perhaps not. Although worldwide oil supplies, according to some experts, may have peaked in 2005, oil's close cousin, coal, is still plentiful—and even dirtier than petroleum. Some geologists a half century ago mistakenly believed that the world would run out of coal within a few decades, a prediction that proved to be wildly wrong. Many nations were later found to have abundant reserves of coal; the United States, India, and China—three of the biggest users—are reckoned to each have a few centuries' worth still beneath the ground. China and India are in fact building inexpensive, coal-fired plants at a breakneck pace to keep up with rocketing demand for electricity. Around 1960 or so, few experts believed that in the 21st century the planet's largest manufacturers would still be heavily reliant on the energy source that kicked off the Industrial Revolution.

To make modern civilization go requires not just energy but a plethora of metals and minerals, from iron to platinum, materials that wind up in skyscrapers, cars, televisions, and computers. For decades the industrialized West has consumed most of this mineral wealth, but a generation ago experts began to wonder what would happen when people in other parts of the globe started buying cars and toasters, too. It was believed that by now some absolutely vital ores would be gone.

In *Limits to Growth*, a report published in the 1970s, experts created charts showing the rate of demand for various metals rising at a few percent a year. Even allowing for advances in mining and the use of alternatives, these projections showed sources rapidly dwindling. By some estimates, tin and silver were supposed to be prohibitively expensive by now; sources of copper and aluminum were to have been completely exhausted.

Paul Ehrlich, not surprisingly, was one of the more vocal proponents of this view, and in 1980, he famously bet economist Julian Simon that in ten years the prices of five metals would rise. When that date rolled around, however, Ehrlich lost: the prices of all five metals had fallen, some far below their 1980 level, a trend that continued up until the early years of the 21st century.

How could it be that there are a lot more appliances, cars, and skyscrapers in the world now than a generation ago, yet we still seem to have enough material to build these things? One example can help explain why. Back in the 1970s, you needed a lot of metal to build a phone system. Millions of people didn't have phones back then, but clearly wanted them, so therefore metal stocks were expected to be plundered to meet this growing demand. But with the invention of the cell phone, poorer countries, especially in Africa, now have phone service without a conventional grid of copper telephone lines. Some of the least technologically advanced places in the world went from having no phones to having the most modern available, skipping the step that would have required billions of dollars in investment.

Prediction: An Inconvenient Goof—the Planet Is Cooling Off

Palm trees in Boston. Malaria cases in Canada. The condos of Florida's coast sitting on the bottom of a bigger, badder Atlantic. Thanks to Al Gore, we all know by now that these are some of the potential consequences of unchecked atmospheric pollution, that if we don't stop spewing carbon dioxide into the sky, we'll bequeath to our kids a world that in general is hotter, stormier, and nastier.

The dwindling supply of climate-change skeptics have a stock reply when confronted with this nightmare scenario— please. Going back just thirty-five years ago, they point to predictions made by scientists who also said we were on the brink of carbon dioxide-fueled doom, but with one key difference— back then, the boffins were saying that the world was getting colder. So which is it, they ask: fire or ice?

For a long time scientists have known that releasing billions of tons of greenhouse gases was going to do something to the atmosphere, but they were hard-pressed to say exactly what. After all, trying to figure out what's going on with the atmosphere is like trying to pick the winner of the Super Bowl at the start of the preseason, except the scientists' task is much more complex. Climatologists must take into account a whole host of variables, from the salinity of the oceans to the vagaries of the sunspot cycle. But sorting everything out has become a lot easier as computers have become more powerful, and today, thanks to machines that can perform billions of calculations a second, a much clearer picture has emerged.

In the 1950s, however, when scientists relied on computers that had just a fraction of the number-crunching ability of a modern-day laptop, it was monumentally difficult to predict the effect man's activity was having on the climate. The result was a schizophrenia that persisted for decades, with scientists figuratively blowing hot and cold.

In 1954, the rough consensus opinion seemed to be that the world had been heating up for about fifty years, but that's where the consensus ended. There was no clear agreement on whether the heating trend would continue, how long it would last, or how abrupt any changes would be.

Back then scientists were already noting that glaciers had been steadily shrinking for several decades—clear evidence of a long-term warming trend, since glaciers, as far as temperature is concerned, are the canaries in the coal mine. But climatologist Hans Ahlmann believed that the upward trend had ceased by about 1950, and predicted that in the coming years temperatures would level off, getting no hotter or colder. Other scientists believed that the warming would continue, but at a snail's pace. Meteorologist George Gamow predicted that the Catskills in upstate New York would become near tropical and that Greenland would be a warm-weather resort—in a thousand years.

The reasons explaining atmospheric change were all over the map, too. Referring to the geological record, scientists noted that temperature appears to be naturally cyclical, with periods where the planet had become warmer and other times when it cooled off. Some experts attributed this phenomenon to the earth wandering in and out of clouds of cosmic dust. Others chalked it up to changes in the amount of heat given off by the sun itself. Still others attributed temperature fluctuations to volcanic activity, noting that the ash and soot emitted during eruptions would deflect sunlight, cooling the planet.

Gradually, however, a small but ever-growing camp came to believe that, first, the earth was indeed heating up, and second, that warming seemed to be occurring in lockstep with mounting quantities of carbon dioxide in the atmosphere emitted from automobiles, factories, and burning forests. In the mid-1950s, Gilbert Plass of Johns Hopkins University was among the first to blame warming on the "greenhouse effect," in which sunlight passes through atmospheric carbon dioxide and hits the earth, which then emits heat. The heat, however, gets trapped by the carbon dioxide, leading to higher temperatures than there would have been otherwise. From the 1950s onward, more and more scientists were coming around to this view, including the prominent physicist Glenn Seaborg, who warned about global warming as far back as 1970.

It took decades for a majority to accept the reality of global warming, however, thanks partly to data that seemed to indicate that winters were getting colder and longer, arctic plants and animals were heading south, and growing seasons were getting shorter. In June 1966, an article in *Science Digest* announced that humanity was "in the dying days of a warming trend." Climatologists back then, charting regular swings in temperature dating back to the 17th century, believed that the planet was emerging from a decades-long hot spell, the Atlantic Ocean seemed to be cooling off, the dry lakes of the American West were expected to revive thanks to increased rainfall, and a general drop in temperature could be expected to last until about 1990, one meteorologist at the Massachusetts Institute of Technology predicted.

By the 1970s, however, some scientists went beyond predicting somewhat frostier winters to forecasting an impending Ice Age, with pretty much all of Canada and possibly even New York City buried beneath a tundra of ice within a few centuries. Others went well beyond even this extreme scenario, believing that an ice age would emerge as soon as the early 21st century.

One of the most notorious—at least to present-day warming skeptics—predictions appeared in the pages of the April 28, 1975, edition of *Newsweek*, which touted on its cover a story about "The Cooling World." The article quoted research showing snow cover at the poles increasing dramatically in the early 1970s and an average temperature drop in the Northern Hemisphere of about half a degree between 1945 and 1968 (which is actually a lot more significant than it may sound to the nonexpert). Within a few decades, the article reported, cold weather would severely crimp the growing season, leading to catastrophic shortfalls in food. Icebreakers would be needed on the Hudson River as far south as New York City.

Contemporary global warming doubters use such predictions to undermine modern concerns about man's effect on the climate. But what these skeptics neglect to mention was that at least as many scientists in the 1970s were worried about warming as about cooling. They also ignore the fact that many scientists, whether "hot" or "cold" earthers, were in agreement on one thing: pollution was having a major impact on the world's temperature. The debate back then centered on which form of pollution was doing the most damage.

Industrial activity produces not just carbon dioxide but soot. As with the effluvia given off during a volcanic eruption, enough dirt in the atmosphere could filter out a substantial portion of the sun's rays, leaving the planet shivering. In fact one modern day scientist—one who believes in global warming—has come up with an emergency plan to save the planet that employs this principle. If we fail to do anything about global warming before things get really nasty, we could send planes loaded with dust and dump the stuff into the upper atmosphere, dropping temperatures just enough to counteract the greenhouse effect—at least in theory.

The point here is that for more than forty years, most scientists have realized that man has proved to be as potent as Mother Nature in affecting the atmosphere. We might not be in the midst of a new Ice Age, but we are rapidly headed toward something that would be just as bad.

Which brings us back to Paul Ehrlich and the other doomsayers we met in this chapter. Like those who foresaw a cooling trend forty years ago, Ehrlich may have been wrong about the particulars, but his general thesis was correct. There has been, and will be, a price to pay for treating the biosphere as though it were an infinitely large garbage can that will nevertheless continue to provide us with everything we need to sustain our wasteful way of life. Anyone tempted to have a chuckle at Ehrlich's expense may come to find that the joke is on him.

Chapter Four

SPACE:
STILL THE FINAL FRONTIER

In 1961, President John F. Kennedy did the government policy equiva-
lent of a baseball player calling his shot at the plate. He predicted that "before this decade [was] out," America would
land men (this was the pre–women's lib era, after all) on the
moon. Kennedy was way ahead of the scientists and engineers
on this one; many of them weren't even sure if it could ever
be done. For starters, a rocket powerful enough to transport
astronauts a quarter of a million miles and back wasn't even
close to being built yet. There were other nagging questions,
too—for example, whether a human being could even survive
an extended stay in outer space, which is awash in radiation
from the sun.

But America pulled it off, and right on schedule to boot, in
July 1969. Given that stunning success and the fact that it was
done in less than a decade, you really can't judge too harshly
the experts forty years ago who thought that by now we would

be spending our vacations on orbiting space stations. Or that thousands of people would be working in factories on the mkoon. And as for Mars? By the early 21st century, of course we would at least have a toehold there.

Some even thought we would already have something like a primitive Starship Enterprise today, a huge, spacefaring craft that could get a good-sized crew to the farther reaches of the solar system. In fact, the original version of *Star Trek*, the 1960s TV show, helped feed this hype with episodes referring to the "early 21st century" as the time when the early forerunners of warp drives and transporters were first developed. A lot of us who grew up in the '60s and '70s, who went to bed wearing our official *Star Trek* jammies, figured that by the time we hit thirty or so, becoming an astronaut would be only slightly more difficult than getting a tractor-trailer license.

A faulty historical analogy also helped convince many of us that space would be a modern playground and factory floor. The astronauts of the 1960s were often described as the 20th-century equivalents of European explorers such as Columbus and Walter Raleigh, intrepid men who paved the way for the millions of immigrants to come. Neil Armstrong's moon landing was going to open the floodgates to massive-scale exploration in the decades ahead—or so it seemed.

The problem with this comparison came down to money. America spent about $20 billion—in 1960s dollars—to get a few dozen men into space and, eventually, on the moon. But it didn't cost nearly as much, relatively speaking, for Spain and Portugal to send explorers to the New World; for one thing, Columbus didn't need a space suit and an oxygen supply to make it to America. Five hundred years ago, building a wooden ship that could carry a few hundred people a few thousand miles was still a lot cheaper than building a space station in orbit today that could accommodate maybe six astronauts at once. And the European exploration—and exploitation—of the Americas soon paid for itself. Europeans got fish, cotton, tobacco, fur, and all sorts of other goodies from their overseas colonies (thanks, in no small measure, to the millions of slaves who also made the crossing). The colonies, especially Australia, were also convenient dumping grounds for criminals. In material terms, though, the Apollo missions yielded just a few hundred pounds of moon rocks—a boon for the scientists but not exactly the same thing as a Spanish galleon loaded to the gunwhales with Aztec gold.

The futurists and sci-fi folks who helped convince us that 2001 was going to see a mission to Jupiter guided by a neurotic computer named HAL basically failed to check with the taxpayer. The average person on the street in the 1960s may have been bowled over by the first photographs of the earth as seen from space, but there were limits on how much people wanted the government to spend simply to feed their sense of wonder. After the moon landing, the American space program didn't die, but it did lose quite a bit of oomph. Americans wanted billions spent here on earth, not to build a fledgling colony in a lunar crater.

So although exploration of the deep reaches of space continues, the *Star Trek* vision of today has pretty much proved to be a bust. The *Star Wars* vision, however, is another story. Manned spaceflight on a large scale may not be economically viable, but space does play an important part in the modern economy—cell phones, GPS systems, and weather forecasters all rely on satellites. So does the U.S. military. Today earth orbit is considered a potential battleground. Destroy enough of a nation's satellite capacity and you could incapacitate the nation—which is why we're spending billions of dollars on weaponry to defend our satellites and, if necessary, get theirs. A lot of Pentagon types also want America to perfect a space-based shield against nuclear attack—an effort that has been under way, off and on, since at least the Reagan era. At this rate, we may wind up with a Death Star long before we have an *Enterprise*.

Prediction: We Don't Have Liftoff

In 1926, Robert Goddard, a Massachusetts college professor, launched the world's first liquid fuel rocket, a gangly contraption that looked like a pointy coffeepot with a huge metal tripod attached. Standing only about twenty feet tall, this humble device paved the way for all the massive rockets to follow, from ICBMs to the Saturn V lunar rocket to today's space shuttle.

At the time, though, Goddard was written off as a crank. No one—at least, no one in his own country—thought much of his little machine. The U.S. military, which today has thousands of rockets in its nuclear arsenal, wasn't much interested. And Goddard's dream of building rockets that could get people into space? Most of the experts back then figured Goddard had been reading too much Jules Verne (and Goddard was, in fact, a huge fan).

People had been building rockets for centuries before Goddard fired up his device in a New England field almost ninety years ago. The Chinese first made fireworks powered by a mixture of charcoal, sulfur, and other substances, sending rockets that were little more than toys arcing through the skies during celebrations. Goddard's innovation involved his choice of fuel. Unlike conventional fireworks, rockets like Goddard's use liquid propellant and a liquid "oxidant," a chemical containing oxygen. Doing this has two advantages: first, liquid rockets are a lot more powerful. Second, because the rocket carries its own oxygen supply, its fuel can burn in space.

Goddard tried repeatedly to get the military excited about his invention, but the generals and admirals of the day apparently believed the only practical use for rockets was on the Fourth of July. One nation, however, did take Goddard seriously—Germany. The Germans immediately grasped the potential for Goddard's device, and, using his work as a starting point, began a rocket-building program for their nation. This would prove calamitous for Allied civilians during World War II, when V-2 rockets loaded with explosives were fired at London, killing hundreds. The machine the American military scoffed at in 1920 was used to deadly effect against its main ally less than twenty years later.

(In an interesting quirk of history, some of the same Germans who built the V-2 were brought to the United States after the war to help develop America's space program. Basically, the Germans carried modern rocketry back to the nation of its birth.)

To the extent that Goddard got any attention at all back then, he was publicly humiliated; this professor of physics was famously accused of knowing almost nothing about the subject he taught. In 1920 an editorial in the *New York Times*, while acknowledging that rockets could fly very high, stated that it was physically impossible for the vehicle to operate outside the atmosphere. "Professor Goddard, with his 'chair' in Clark College . . . does not know the relation of action to reaction and of the need to have something better than a vacuum against which to react. . . . Of course he only seems to lack the knowledge ladled out daily in high schools." (In an equally famous mea culpa, the *Times* later corrected that editorial—a few days after men landed on the moon in 1969 using a rocket that worked just as Goddard predicted it would.)

By the 1950s, rockets capable of leaving the atmosphere had indeed been built, leading many to conclude that humanity was now very close to realizing the goal of traveling in space and even to the moon. Yet still there were skeptics. Lee de Forest, the radio pioneer, said in 1957 that a lunar journey was a "wild dream": "I am bold enough to say that such a man-made voyage will never occur regardless of all future advances." Other critics were more succinct. "Space travel is utter bilge," Richard van der Riet Wooley proclaimed when he became British Astronomer Royal in 1956.

Even some of those willing to concede the possibilities underestimated the speed with which they would be realized. A Minneapolis researcher, Otto C. Winzen, predicted in the mid-1950s that men would travel in outer space no sooner than 1985 (the Russian Yuri Gagarin went on his historic flight in 1961). Writing around the same time, the scientist A. M. Low believed it would be another century before man set foot on the moon. "Columbus . . . had the considerable knowledge of sea navigation accumulated during several hundred years. We know nothing of space navigation, and of what lies beyond about 20 miles of the earth's surface we are almost entirely ignorant."

Within just a few years, as the Soviets and Americans sent dozens of men into low earth orbit and, eventually, to the Moon. Low, van der Riet Wooley and the other doubters were proved wrong. As a result, prevailing wisdom in the 1960s held that a lunar landing was to be just the first, and least ambitious, of many voyages to follow.

Prediction: We Have a Base in Space

John F. Kennedy announced that America would land on the moon within ten years; Americans were inspired. More than forty years later, when George W. Bush announced that NASA was going to send people to Mars, Americans were amused. During the last fifty years, manned, interplanetary exploration went from being a noble national goal to resting somewhere near the bottom of the nation's to-do list.

It would have shocked a lot of people back in the 1960s that not only had we not landed on yet another planet by now, but we weren't even close to doing so. No one has set foot on another world since the end of the *Apollo* lunar missions in the early 1970s. In fact, many experts today have serious doubts about the usefulness of manned, interplanetary space flight at all. Even in the '60s advanced robot explorers were doing a pretty good job of mapping the rest of the solar system, and the probes are even better today. In a sense, anything an astronaut can do a machine can do better (and more cheaply).

But decades ago, the only justification needed for sending people on Martian voyages was the same one heard for scaling Mount Everest—because it's there. Space exploration helped raise America's prestige and, nearly as importantly, was considered cool to a generation being raised on *Star Trek*. If you ever hoped to meet Vulcans and Klingons, you had to start somewhere.

All this was enough to motivate some of the best minds of the time to draw up plans for manned missions to our nearest planetary neighbors, Mars and Venus. These engineers and scientists had no illusions about the scope of the challenge they had set for themselves. But thanks to their considerable ingenuity, they devised plans that were at least doable—provided the political will was there.

At its closest, Mars is about fifty million miles away, so engineers thought a trip would take about six hundred days, plus whatever time was actually spent on the planet. Many scenarios envisioned astronauts spending quite a spell on Mars, first for exploration purposes, and second to restock vital supplies of air and water. This last part was crucial. One of the main challenges behind a manned Martian mission was devising rockets powerful enough to transport not just the people but the thousands of pounds of supplies necessary to sustain them for the trip. Using Mars as a supply post would cut down on how much the astronauts would have to bring with them.

In 1955, Robert Richardson, writing in the *Saturday Review*, envisioned a crew setting up a base near one of the Martian poles, where there was believed to be a lot of ice. The poles could be tapped for water and, through a simple chemical process, could yield oxygen as well (which is in short supply in the Martian atmosphere). Richardson wrote that the astronauts would have to follow a strict regime, spending nearly all their time in a base that was as cramped as a nuclear submarine.

A decade later, after much more had been learned about space flight, famed rocket expert Wernher von Braun believed—along with dozens of other scientists—that a manned Mars landing would occur well before the 21st century following a series of intermediate steps. By the mid-1970s, it was thought, a small group of humans would spend considerable time in orbit so more could be learned about the effects of zero gravity on the human body—a prediction that proved more or less accurate. For the last several years, astronauts have spent months at a time aboard the International Space Station, the largest man-made object revolving around the earth. One astronaut was even married there.

But low earth orbit is as far as human beings have gotten since the space race days of the late '60s and early '70s.

While von Braun was right about humans spending time in low orbit, he was pretty off when it came to his timetable for a manned mission to Mars. Forty years ago, von Braun predicted, a small crew would be sent to Mars orbit around 1980, a strictly reconnoitering mission that would set the stage for a future landing (the earlier *Apollo* lunar flights also involved sending astronauts around the moon first before the landing was attempted). A few years later, in the mid-1980s, man would finally set foot on the Red Planet, Von Braun thought.

How would we get there? Most scientists realized then as they do now that conventional rocket fuels would not do the trick, because they don't deliver enough thrust to propel a craft weighing tens of thousands of pounds on a journey of tens of millions of miles. Von Braun and others believed that some variety of nuclear propulsion would have to be developed. The simplest approach involved controlled nuclear explosions, in effect directing the force of the A-bomb through some sort of nozzle at the back of a craft. Others believed that lasers, powered by solar panels, could heat up a propulsive gas to send the ship on its way.

Then there was the "solar sail," a system that would be charmingly reminiscent of the schooners and galleons of old.

This ship would use a massive sheet of thin metal that would collect the energy that constantly streams from the sun, the so-called solar wind. To cut down on the amount of food and water required for such a lengthy voyage (a sail-powered craft would take years to reach its destination), astronauts would spend the bulk of the trip in suspended animation, a state in which all the bodily processes would be slowed to a point just short of death.

None of this was going to be cheap. While a few projections indicated that a Mars mission could be achieved for about $100 billion (spent over many years), von Braun and others put the price tag much higher: about 1 percent of the nation's annual gross domestic product. Hard as it is to imagine today, back then von Braunn believed that Americans could be persuaded to spend one dollar out of every hundred the economy created strictly for the sheer joy of discovery.

Von Braun wasn't the only one who thought so, either. In the late 1960s, Arthur C. Clarke, the highly respected writer (rare among science fiction authors, Clarke had also done pioneering scientific research), described a trip to Mars as likely to happen by 2000, as did the futurists Wiener and Kahn. A decade later, in 1979, a poll conducted by *Omni* magazine revealed that most readers believed that there would be a manned mission to Mars by 1992. Only 3 percent—the lowest rate among several questions *Omni* posed—believed that a Mars mission would "never" happen.

Even if a Mars trip proved too expensive, or just too hard, to pull off, it was taken as an article of faith at one point that we would have lunar bases in the early 21st century. After all, people actually had been to the moon. A group of scientists at TRW, a defense contractor, gave 1980—just thirteen years later from when they wrote—as the time by which the first lunar base had been established.

Around 1970, Isaac Asimov, another acclaimed scientist–science fiction writer, said that within fifty years—by about 2020 or so—thousands of people could be living in underground lunar caves. Water would be retrieved from deposits frozen in the lunar crust. Electricity would come from solar power—an especially likely prospect, because one side of the moon is perpetually facing the sun and there are never any cloudy days. Oxygen would come from recycled air breathed out by the moon residents; water and human waste would likewise be reused. All of this would be especially likely, Asimov believed, if the United States stopped competing with the other great space power, the Soviet Union, and started working with it.

Given that the Russians were also thinking of exploiting the moon back then, this made a lot of sense. Nikolai Kolomiets, writing in the publication *Red Star* in 1967, envisioned domed cities where plants and flowers bloomed beneath the eternal sunlight. Dwellings would be blasted out of the lunar rock. Kolomeits, noting that such "micro-cities" were already being constructed in Antarctica and in Russia's equally inhospitable far north, felt confident that similar encampments would also work in the harsh environment of the moon.

Other proposals were even more far-fetched. In the late 1970s, Carolyn Dry, an architect who specialized in using materials to hand in erecting buildings, believed that cheap but durable glass domes could be erected on the moon using rock already there. Under these domes would grow algae, plant life that exhales oxygen, and eventually the algae would be let loose on the lunar surface, leading to the creation of a primitive atmosphere. This is roughly the same process whereby earth acquired its atmosphere, Dry noted. Some even predicted that, employing these or similar methods, a lunar colony would become so well established that the first moon baby would be born right around the year 2000.

There's a chance that some country or group of countries might go back to the moon within the next few decades; the Chinese, in particular, seem very interested in planting their flag on the dusty lunar surface. But if, or when, a lunar base—much less a lunar maternity ward—will be built is difficult to say.

Prediction: The Orbiting Time-Share

Baby boomers who saw the 1968 film *2001: A Space Odyssey* were enthralled both by the power of Stanley Kubrick's filmmaking as well as the promise for the future implicit throughout. In one early scene, for example, a group of passengers is ferried to a giant, slowly rotating space station via a rocket that resembles a jet airliner—operated by TWA, no less, a real-life airline that, presumably, had branched out by the time the 21st century rolled around. The suggestion was clear: space travel would one day be almost as commonplace as flying.

Most of the predictions in this book were not the product of a sci-fi author's brain, since, by definition, the scenarios in this kind of literature were not intended as literal forecasts of what was to come. In the case of *2001*, however, an exception should be made. The film was based on the book by Arthur C. Clarke, who took great pains to ensure that the details were technically accurate. Clarke, who in the late 1940s first described how communications satellites would work, knew what he was talking about. If NASA's drawing boards were any guide, both routine space travel and massive space stations seemed less than three decades away—something that Clarke was aware of.

What was needed to realize this vision was a ship that could economically transport hundreds of "commuters" to low earth orbit—and for decades now, such "space planes" have been in the very early development stages. The most common design is known as the "scramjet," a vehicle that is part rocket, part airplane. Since at least the 1950s, space planes employing this propulsion system have been in the works, but they have always faced some severe obstacles. A craft that flies at Mach 24—fast enough to go into orbit—would experience tremendous heat and pressure. To do the necessary testing of such a machine requires incredibly powerful computers, which, until fairly recently, had not been built. But even with the advent of such machines the rocket plane still doesn't exist outside the imagination.

To develop a rocket plane would require billions of dollars in investment, and no major airplane manufacturer is sure if the market is there. But decades ago it was believed that there would be plenty of reasons to go into space, where there was supposed to be all manner of economic activity—a kind of Main Street a few thousand miles up.

In 1967, Barron Hilton, the scion of the famous hotel family, said he would build a resort in space if freight costs became sufficiently cheap—about fifteen dollars a pound, by his reckoning. Assuming such low freight costs, a lot of other space businesses were in reach, too. Futurist Theodore Gordon, writing in the late 1960s, envisioned hazardous materials being shot toward the sun aboard inexpensive rockets. Stroke and heart patients would spend a recuperative spell aloft in stress-free zero gravity.

A lot of manufacturing was supposed to be happening in space by now, with goods exported back to earth in what might be described as extra-globalization. Many industrial processes work best in low gravity or in low temperatures, such as the production of electronics. Writing in *Fortune* magazine in 1979, Gene Bylinsky, an author who wrote about space exploration, said that goods could be produced in orbit that had a value of about thirty thousand dollars a pound, a bonanza so rich that "no corporation affected by changes in technology can afford to ignore the new era of innovation that is about to begin."

A lot of this manufacturing was to happen on the lunar surface. Neil Ruzic, in his 1965 book *The Case for Going to the Moon*, described mining operations set up to exploit the billions of tons of uranium, cobalt, and other precious metals to be found there. Because of the feeble gravity, ore and products could be launched from the lunar surface relatively easily using a variety of methods, including an electromagnetic high-speed rail that would fling containers laden with cargo toward the earth. Ruzic thought that these nuclear- or solar-powered facilities could be up and running by about 1985 or so.

Space was supposed to provide another kind of bounty, too: energy. As far back as the 1950s engineers envisioned giant, space-borne solar collectors converting sunlight into electricity and then beaming it back to earth, providing a limitless supply of power. Although the details changed over time, such a space generator would literally remake civilization, at a stroke ending debates about fossil fuels or what to do about nuclear waste.

Yet another far-out scenario involved the creation of an artificial "moon," like the solar collector a giant parabolic mir-

ror set into orbit. Using reflected sunlight, the mirror would
be positioned to provide a light source for selected regions of
the planet at night. Some believed that doing so would enable farmers to grow even more crops in a perpetual daytime.
Cities in far northern latitudes could find some relief from
depression-inducing twenty-four-hour nights.

Perhaps the most ambitious plan of all was the construction of an entire city in orbit, a slowly rotating community that
could house thousands of people. First proposed by Princeton
University professor Gerard K. O'Neill in the mid-1970s, such
a floating city would consist of two concentric tubes, resembling bicycle wheels, whose rotation would supply inhabitants
with gravity. Sunlight would nurture plants, which in turn
would provide oxygen for the thousands of inhabitants, creating a closed, self-sufficient ecosystem. O'Neill intended this
mega–space station as the forerunner for a kind of lifeboat for
the human race. Since he, like many others at the time, was
concerned that the earth would no longer be able to sustain
rapidly multiplying humanity, "re-creating" earth in space
seemed like a viable option.

Few believed that O'Neill's space city would have been
built by 2008, but many thought we would be further along
than we actually are. Far from having thousands, or even hundreds, of people in orbit at any given time, however, there
aren't enough souls floating above our heads today to fill a
grade school classroom. Nor is the industrialization of space
nearly as advanced as many thought it would be by now.

To build space cities, hotels, or factories would have
required, as Hilton had correctly pointed out, a cheap way
to get thousands of tons of construction materials into orbit from earth. Recognizing this, engineers at NASA and

the major aerospace companies designed relatively inexpensive, almost disposable "space trucks," mass-produced rockets simple enough to be launched hundreds of times a year.

But these basic designs, dating from the mid-1960s, gradually evolved into the space shuttle, a temperamental machine that was completed in the early 1980s after years of hard work. Because the shuttle is so complex, launches are neither routine nor cheap. The shuttle program suffered two dramatic, tragic setbacks when the *Columbia* and the *Challenger* exploded in 1986 and 2003, respectively. Each time shuttle launches were delayed for long periods of time as engineers labored to iron out the bugs, preventing yet more tragedies. Slated to end in 2010, the shuttle program, on balance, has been labeled a disappointment by many commentators.

That's not to say that space has not been commercialized to some extent. Entrepreneur Richard Branson has launched Virgin Galactic, a service that will bring paying (and well-heeled; a ticket costs about two hundred thousand dollars) customers to low earth orbit for a brief but thrilling ride. Jeff Bezos, the founder of Amazon, is trying to develop a space station using relatively simple, inexpensive modular construction techniques. In 2004, Burt Rutan won the $10 million "X Prize," which was given to the first private entity that managed to launch a vehicle into space.

Other entrepreneurs are also working on vehicles to deliver payloads and people into orbit, as well as to create space stations. In fact, private citizens such as Bezos and Branson think they are better suited to the task than huge government agencies such as NASA. The profit motive may prove to be the most powerful boost to space exploration ever known.

Prediction: Star Wars—for Real

Life on earth as we know it today would be a lot different if
we did not have a presence in space. Cellular phone communication and global positioning systems require satellites.
Satellites are also used to help farmers measure crop changes
and by engineers to design roadways, dams, and other structures. The average person can now go online and get virtually
real-time, close-up images of nearly any spot on the globe—
technology that was available only to the military just a few
years ago.

The modern military, in fact, does rely on space as much as
the average, modern-day person does—perhaps even more so.
The navies, armies, and air forces of major nations are heavily dependent on satellite communication, space-based navigation, and eye-in-the-sky reconnaissance. We may be close
to the day when we'll be placing actual weapons—whether
offensive or defensive—thousands of miles above the earth's
surface. While some forecasts have not come to pass—at least,
not yet—many planners were correct in predicting that earth
orbit would be an increasingly vital battleground in the 21st
century. With the world's most advanced military, America
is especially vulnerable to attack against its space satellites. A
potential enemy could effectively shut down America's entire
communications system. In the last forty years, the region just
outside the atmosphere has become another potential front
in war.

That's why the United States has been observing the rise
of China's space program with more than a casual interest,
especially after the Chinese successfully tested a missile that
brought down a defunct satellite a few years ago. Although the
Chinese denied that the test was intended to send a message,

Western military planners could not help but take heed. Any nation that had the capacity to disrupt space-based defenses had to be taken seriously. An effective attack against a country's military "assets" in orbit would leave that country blind and mute—and highly vulnerable.

Long before the first astronauts cavorted in zero gravity, military planners viewed low earth orbit as the ultimate high ground; any nation that owned space, it was believed, would enjoy a virtually unbeatable military advantage. The rockets powerful enough to leave earth's atmosphere were developed in the years immediately after World War II, around the same time that the United States and the Soviet Union were also developing their nuclear arsenals. In an era when schoolchildren were being taught to hide under their desks at the first wail of the air raid siren, it was believed that the difference between the winner and loser of an atomic war was just a few minutes. Survival depended on a country's ability to drop its nukes on an enemy before that enemy could unleash its own bombs.

Space was therefore not seen as an arena for heroic exploration but rather as vital territory that had to be captured, a case made by von Braun in the early 1950s. With the Cold War well under way, von Braun proposed building space platforms that would house nuclear missiles. These platforms, coupled with advanced monitoring equipment, would allow the United States to bomb an unruly foe before that enemy had a chance to do much damage. Another version of this idea called for missile installations on the moon, sites that would be invulnerable to all but the most sophisticated attack and manned by a large crews of technicians—a kind of lunar NORAD.

Von Braun, somewhat naïvely, believed that such an arrangement would lead to world peace—so long as a benign nation such as America held the ultimate trump card, nations would no longer fight one another. America would, as a later generation put it, become the world's police officer, an omnipotent cop in the sky.

The great, obvious flaw in von Braun's plan was that for every move the United States made, the Soviet Union and other enemies would certainly counter it. If the technology existed to create space-based missile launching pads, then the capability to take those launching sites out would, of course, exist, too. The Russians would no doubt put their own complement of warheads in space. Far from leading to world peace, any nation that put missiles in orbit would only manage to extend the arms race into new territory.

Following the Cuban Missile Crisis in 1962, a wake-up call to both the Cold War protagonists, talk of using space as a haven for nuclear weapons evaporated. Treaties signed by Russia and America in 1963 and 1967 explicitly banned placing weapons of mass destruction in orbit or on the moon, and for the next twenty years both nations focused on building ever more advanced observation and communications satellites. But as those satellites became more sophisticated, and hence more militarily valuable, space was once again opening up as the site of a possible shooting war.

Writing in the early 1980s, science fiction writer and space theorist Ben Bova described a scenario in which Russian satellites, armed with conventional missiles, would fire upon the then-new space shuttle, damaging it. Cosmonauts flying in from a manned space station would then commandeer the shuttle, repair it, and fly it back to the Soviet Union, on the grounds that the shuttle was a military craft being used to threaten Soviet satellites. Bova also imagined Russia's orbiting gunboats taking shots at enemy—that is, American—satellites that drifted too close to "Russian" orbits.

Furthermore, Bova noted that the Soviet Union had been working on a type of space mine, a satellite loaded with explosives that could be parked into orbit next to a target and detonated. He even suggested that such a weapon may have been turned on a U.S. communications satellite that had gone missing in 1979.

But this was child's play compared to the "beam" weapon, a device that was then being tested by both the United States and the Soviet Union. Satellites equipped with powerful new lasers or guns that could shoot beams of charged particles toward a target could punch holes into the fuselage of a nuclear missile from several miles away. If such a weapon had been developed, it could enable one country to destroy another's incoming nuclear warheads before they reached their target. Since the best offense is a good defense, beam weapons would tilt the balance of power toward whoever possessed them. One writer, Jerry Pournelle, said in the early 1980s that space-based battle platforms would definitely be in place by the 21st century, and probably by 1990. What's more, Pournelle was convinced, those weapons would be in Soviet hands, thanks to the seemingly huge technological lead the Russians had in developing them. The nuclear stalemate that had been in place for decades would finally be broken—and America would be the loser.

Soon after, the creation of so-called Star Wars weapons would leave the realm of speculation and enter the realm of official U.S. policy. In a famous speech in 1983, President

Ronald Reagan announced the Strategic Defense Initiative (SDI), a plan to put laser platforms and hyperaccurate guns in orbit (a plan also endorsed by the U.S. government's General Accounting Office). Critics howled, pointing out that such space-based defenses were probably a violation of the 1972 treaty prohibiting the development of antiballistic missile weapons. Others believed the whole thing was just a multibillion-dollar boondoggle. Reagan, however, remained undeterred: "We're talking about a defensive shield that won't hurt people but will knock down nuclear weapons before they can hurt people," he said in 1985.

But the SDI was put on the shelf with the collapse of the Soviet Union a few years later. Although some politicians still wanted to see these weapons developed—creating a space-based missile defense system was a little-known part of the Republican Contract with America in 1994—the concept pretty much languished until the September 11, 2001, terrorist attacks against America.

More than one commentator has pointed out that modern "rogue" nations pose even more of a threat than the Soviet Union ever did, primarily because the Soviets, no matter how bellicose their rhetoric, did not want to die. Today, however, there are groups willing and eager to give their lives for their god—groups that could conceivably seize control of a nation armed with nuclear missiles, many security experts fear. North Korea, which has no interest in carrying out bloody jihad, nevertheless is led by a man, Kim Jong Il, who has already proved his willingness to let hundreds of thousands of people starve rather than surrender to outside demands. With total control over his country and armed with long-range missiles as well as nuclear bombs, Kim alone justifies the creation of an elaborate U.S. defense system in space, some feel.

Today, America has stepped up testing of antimissile particle beam and laser weapons (even though North Korea presently can't lob missiles powerful enough to go into space, beam weapons are seen as a hedge against the time when they, or another equally hostile nation, will be able to pull it off). A few years ago, the then-secretary of defense, Donald Rumsfeld, endorsed the idea of putting such weapons systems in orbit. At the same time, both the Chinese and Americans are developing rockets, launched from high-flying jet aircraft, that could destroy orbiting satellites. Sadly, it seems likely that there will be a Death Star long before there is a Hilton in the heavens.

Chapter Five

NUCLEAR FUSION, WEATHER CONTROL, AND OTHER TECHNICAL MARVELS

In our culture we used to regard technological advances the same way we did our high school crushes: we loved 'em blindly. In the middle decades of the 20th century, science was taking down major diseases like polio, making food abundant and a lot cheaper, and giving us exciting new diversions, such as TV sets and transistor radios. No one really thought much about any downside, though; we were focused mostly on how these achievements could make our lives better. True, no one was much thrilled with the arrival of the nuclear bomb, but people tended to zero in on how atomic energy would lead to cheap electricity. We saw past the clunky braces, cracking voices, and bad skin and embraced the best of what the laboratories and engineers had to offer.

This was a time when Americans seemed to dream bigger. We built the tallest skyscapers, the biggest dams, and a highway system that stretched across the three-thousand-mile breadth of the country. We cooked up plans to make the deserts bloom and the mountains tremble. We would make the rivers go exactly where we wanted, not where random, flighty Nature decided. Although no one said so out loud, many thinkers back then believed we could, in effect, do almost anything we desired, and without any consequences.

But beginning in the early 1960s, some started to wonder whether remodeling the planet in any way we saw fit was really such a good idea. We were gradually becoming aware that the technology that could improve our lives in one way could also make them a lot worse in other ways. In *Silent Spring*, Rachel Carson wrote about how the very pesticides that made farming more efficient by killing pests could also eventually kill us, too. More and more people began to worry about the countless birds, plants, and animals that were lost whenever we built a new subdivision. Jet travel was great, but nobody especially wanted to live beneath a flight pattern. The factories that produced our toasters and cars also belched out the smog that gagged us and smarted our eyes. By 1969, when the pollutants in an Ohio river spontaneously caught fire, our love affair with technology hadn't ended, but it had matured. Today an army of lawyers remains constantly on the lookout for any ill effects that might come along with the next great technological advance. A slew of environmental regulations serve as a check on the ambitions of the wild-eyed dreamers in the lab coats. In short, we're still married to our technological world—but now we insist upon a prenup.

Prediction: The Atom Keeps It Together

It sounds incredible, even today: just a few pounds of uranium are enough to level a city, power a battleship, or electrify a thousand homes. So incredible, in fact, that when nuclear energy's potential was first discussed publicly seventy or eighty years ago, a number of experts—including some of the most acclaimed—weren't buying it.

In the late 19th century, scientists first began to understand the composition of one of nature's fundamental building blocks, the atomic nucleus. The atom consists of protons and neutrons locked together in a core surrounded by orbiting electrons. They also learned that the force needed to keep the nucleus's components bound up was incredibly strong—and if it were unleashed, you would have a weapon more powerful by far than anything that had come before. And if the atom could be smashed apart under controlled conditions, you would have the fuel for a new, unprecedented type of power—nuclear power.

Many researchers, Albert Einstein among them, believed early on that we would harness the power of the atom, for better and worse. But the idea of superbombs capable of destroying New York or of using just a dash of uranium to run a battleship led to a lot of skepticism, too. In the early 1920s, for instance, Scottish scientist J. B. S. Haldane wrote that he did not "much believe in the commercial possibility of induced radioactivity." A decade later, Yale professor C. C. Furnas, despite advances in atomic physics, cautioned the public "not to buy stock in an Atomic Energy Development Company. You will certainly lose." Scientists quoted in a 1931 edition of *Science Monthly* described the possibility of atomic power as "nonsensical."

The most stunning prediction, however, came from Ernest Rutherford, a New Zealander born in 1871 who is recognized as one of the founding fathers of atomic physics. Rutherford predicted the existence of the neutron, helped build the first Geiger counter, and was the first to accurately describe exactly what radiation is—a slow disintegration of atomic particles. When it came to the possibilities of the atom, you couldn't ask for a better source than Rutherford—or so it would seem.

Throughout the 1930s, Rutherford mocked the young physicists who said that the atomic age would soon be upon us. In 1933, he said flatly that we would never derive power from the atom. Rutherford knew full well the energy potential locked up in the tiny atomic nucleus, but he also believed that you would have to employ more power to crack the atom open than would be released after it was split. Yet in 1942, just five years after Rutherford died—and he went to his grave an atomic energy skeptic—the world's first controlled nuclear chain reaction occurred at the University of Chicago. The energy from such chain reactions heats water that drives steam turbines in today's nuclear power plants.

By then World War II was under way, and the achievement at Chicago was a milestone in the secret Manhattan Project to build an atomic bomb. The reaction in an atomic bomb explosion is similar to that used in generating electricity, except that in the bomb a lot more heat is released much more quickly. The bombs then being designed would each have the explosive force of thousands of tons of TNT—so powerful, in fact, that physicists working on the project worried that the bomb's detonation might crack the earth's crust or set the atmosphere on fire.

Others, however, were only concerned that the United States was wasting a lot of time, money, and effort. "This is the biggest fool thing we have ever done. . . . The bomb will never go off, and I speak as an expert in explosives," Admiral William Leahy wrote to President Harry Truman—the same president who would soon afterward approve the use of atomic bombs against Hiroshima and Nagasaki, Japan.

For the next twenty or thirty years, almost as if to atone for atomic energy's ghastly debut, many commentators predicted that the nuclear power plant would lead to a new golden age. According to various estimates, anywhere from half to all of America's power was to come from nuclear plants by the 21st century (the United States today only gets about a fifth of its power from the atom, although the percentage in France is much higher). Writing in 1966, the respected *New York Times* science correspondent Walter Sullivan quoted a researcher at the Tennessee Valley Authority who believed that atomic energy would soon be dirt cheap, making fantastic new industrial processes economically feasible. The cost of manufacturing everything from fertilizers to steel would drop by half or more thanks to this inexpensive new energy source. Around the same time, another commentator said that "it could not seriously be doubted" that America would be getting all of its electricity from nuclear power by now.

These broad pronouncements for our nuclear future were made, of course, long before Three Mile Island and Chernobyl. It was also before the builders of atomic energy plants discovered that getting a permit to construct a plant could take years (many industrialized nations, including America, have gone decades without a new nuclear plant coming on line for this very reason). In addition, this was an era before the general public had really learned to hate nuclear waste; as of 2008, lawmakers in Nevada were still opposed to the facility at Yucca Mountain, where the Atomic Energy Commission would like to store old nuclear plant components that will remain radioactive for thousands of years. It was, in short, before the power of the NIMBY—not in my backyard—phenomenon was fully appreciated.

Today, nuclear power is seemingly back in fashion again. In the United States and in other countries, companies are seeking to build the first new plants in years, and even some environmentalists would rather see us get more power from emissions-free nuclear plants than dirty coal. But the fundamental problems still haven't been licked. No one really wants to live in the shadow of the cooling towers, and no one has yet figured out what to do with waste that will remain radioactive for ten thousand years. Energy that was once billed as being "too cheap to meter" still comes at a price.

Prediction: The Nuke as Steam Shovel

Ever since the dawn of the nuclear age, we've been living in fear that one day, hundreds of nuclear bombs would be set off over our soil. But back in the 1950s, engineers believed we would be uncorking mushroom clouds all the time—and on purpose. Whenever we wanted to widen a canal or build a tunnel, some once thought, we would set off the bomb.

Under an initiative called "Project Plowshare"—as in what a sword gets converted into—the U.S. government detonated a nuclear bomb in Nevada in the early 1960s in order to create a crater and to sell builders on the idea of using nuclear weapons in construction projects. The idea was to show that nuclear bombs could not only destroy, but also create; that the immense power of the atom could be harnessed to remodel the earth in any way we see fit.

Mining and drilling for oil seemed to be tasks particularly suited for nukes. For a long time, the world's oil companies have known of deposits that, in theory, would be worth a fortune, and yet they left those billions of barrels untouched. Finding the oil is only half the battle—sometimes not even half. What can be even trickier is getting the stuff to the surface at a reasonable cost. Even during the oil shock era, geologists, probably slamming down their charts in frustration, realized that not every mother lode is like that found in Saudi Arabia, whose black gold is precious thanks to the fact that it is relatively easy to get to.

Setting off a nuclear bomb beneath the surface is one way around this difficulty. In the 1950s, '60s, and '70s, geologists had discovered deposits deep underground, often wedged between layers of rock that could not be breached using conventional drilling techniques. A well-placed atomic warhead, however, could easily shatter the rock in which this oil was trapped.

There was something of a precedent for this. In the '50s, the United States and Russia used to conduct aboveground tests of nuclear weapons in remote areas, inadvertently spawning a slew of B-films, such as 1954's *Them!*, starring hideous creatures that had been mutated by clouds of noxious fallout. Military scientists on both sides soon realized that releasing tons of radioactive dust in the atmosphere was probably not such a good idea (although there was little risk of creating ants as big as houses and other Saturday matinee horrors), so the testing went underground.

Although that practice too eventually fell out of favor, scientists had nonetheless learned that you could in fact set off a nuclear bomb below the surface safely. Writing in the 1960s, a Brussels-based researcher believed that atomic mining would be commonplace by the early 1980s, provided that public opposition could be overcome.

That seemed a lot more likely with the oil shocks of the 1970s, when skyrocketing fuel prices hit the average person right in the wallet. A poll of energy experts in 1974 revealed that most believed nuclear explosions would soon be routinely used for the production of oil. Around the same time Russian scientists, noting that about 40 percent of the world's petroleum was trapped between layers of impermeable rock, had started conducting tests to extract oil and natural gas using this method. The problem, as one of those scientists noted, was finding sites where nuclear bombs could be set off without contaminating groundwater supplies.

In the following years, the idea gradually faded away, as new extraction techniques, such as horizontal drilling, were developed. But at the same time, civil engineers believed the atomic explosion had other, nonmilitary uses as well.

In the mid-1950s the physicist Edward Teller proposed using nuclear weapons to build a replacement for the Suez Canal, which had been taken over by the Egyptian government. (Teller also wanted to drop a nuke on the moon—just to see what it was made of.) By the mid-1960s there were plans to create an artificial harbor in Alaska using A-bombs. That idea never got off the ground, however, because of environmental impact studies that revealed the obvious—no ship would ever be able to safely dock there, thanks to lingering radiation. The area where the test was supposed to take place was sparsely populated, but the people who did live there were not exactly thrilled with the idea of having their homes irradiated for the sake of progress.

The same problem killed a 1967 plan to build a second Panama Canal using a few nukes in place of thousands of malaria-infected ditch diggers like those who excavated the first canal. A site report for the project, apparently never completed, was due in 1970. Under the plan, hundreds of bombs were to be set off underground, creating a second canal a lot more cheaply than the first one had been built. It was an ingenious idea—if but for that pesky radioactive contamination.

Prediction: I Can Build Ya an Ocean

Given the scope of progress that the 20th century witnessed, it should come as no surprise that there was a lot of confidence that we'd be able to engineer a better natural world for ourselves. For centuries we'd been utilizing our tools and ingenuity to harness nature and control it. Using slave labor and tools no more sophisticated than the pulley, mankind built the pyramids and the Great Wall of China. Aqueducts that cut through mountains and spanned rivers brought water from hundreds of miles away to 1st-century Rome thanks to straightforward innovations like the arch and a good formula for concrete. Medieval Dutch engineers managed to push back the sea with simple earthen dikes. Engineers in the late 19th and 20th centuries, armed with an even bigger and more sophisticated toolbox, became heady at the thought of what they could accomplish.

The middle of the last century saw something of a high-water mark for these reveries. For example, the Futurama exhibit at the 1964 World's Fair included a description of a massive, automated, atomic-powered (of course!) road-building machine that was designed to operate in the jungles of the Amazon. Running around the clock, the device, which was about the size of a freight train, would use lasers to hack through the dense underbrush and bucket loaders to scoop and grade the exposed soil, while macadam would squirt out the bottom. Blacktop would be laid down at the stunning rate of twenty miles a day, and far less expensively than if human beings were doing most of the work. Never mind that setting such a machine loose in the Amazon would also destroy God knew how many specimens of rare plants and animals; that was not a primary concern at the time for General Motors, the exhibit's sponsors. Instead, seeking only to maintain worldwide demand for Chevys and Cadillacs, America's premier car company was doing everything it could to encourage massive-scale road construction, applying a template globally that had worked so well in its home market (for decades, America's carmakers assisted lobbyists seeking more government money for highways).

This was a time when, clearly, the Amazon rain forest got no respect, and sending a nuclear-powered weed-whacker into the pristine region now known as the planet's "lungs"—the forest produces a lot of oxygen—wasn't even the worst of it. A few years after the Futurama exhibit closed down, the Hudson Institute, an American think tank, proposed creating an inland ocean the size of Germany by damming up the Amazon and other South American rivers. As part of the project, engineers would also build six additional bodies of water, each the size of Lake Ontario, that would be sprinkled throughout the continent.

The idea was to create a continuous waterway link across the fat part of South America connecting the Atlantic and

Pacific oceans. The South American interior is low-lying and once did contain a huge body of water, so the topography was well suited for the project, its proponents noted. The man-made waterway would make the timber- and mineral-rich inland region easily accessible to ports (the trade-off, of course, was that thousands of square miles of "worthless" forest would be flooded). All those dams would also mean plentiful hydroelectricity for a part of the world that was then casting about for inexpensive sources of power as Colombia and other nations industrialized. The project also called for an invitingly cheap, low-tech approach: the dams would be simple earthen mounds a few hundred feet high, and the final tally would be something like a billion dollars. That sum was fairly modest considering the return that could be realized once the jungles of South America became open to easy exploitation.

Although there were some concerns—one French engineer speculated that placing such a huge weight of water so close to the equator could actually slow the earth's rotation by a few seconds a year—the South Americans were seriously entertaining the concept. Colombia, Bolivia, Peru, and Brazil all formed commissions to examine the idea. Governments looked for groups of private investors. Scale models were built and hydrological studies conducted. Robert Panero, an engineer who did much of the preliminary work on the proposal, noted with some satisfaction that the South American ocean would be the only man-made object, aside from the Great Wall of China, that would be visible from space.

Not to be outdone by their Cold War nemesis, the Russians had similar ideas. Throughout the 1960s there were plans to create an inland ocean between Lake Baikal to the east and the Aral Sea to the west by damming the Ob and Yenisey rivers, which flowed into the Arctic Ocean. Water from this man-made sea would then be conveyed south to the dry steppe, opening up more than sixty million additional acres to agriculture. Not only did the Soviet Union fail to achieve this, but planners actually managed something quite the opposite. Thanks to intensive irrigation, the Aral Sea is now a lot smaller than it was when engineers first devised their plan to create a new ocean in the heart of Russia.

Crazy and ambitious though it may sound, building inland oceans at least had a precedent; early in the 20th century, thanks to a dam break, the Salton Sea was formed in a Southern California basin where there were once extensive salt mines. But other plans to mimic nature were wholly novel—such as a proposal to create an artificial volcano.

In the mid-1960s, writer Dennis Gabor described a plan in which machines would burrow down to the magma layer in Tierra del Fuego or some other volcanic region using an electric drilling method. Once the magma layer was reached, water and air would be pumped down, creating a dense steam laden with droplets of iron that would rocket upward back toward the surface. The residue would be collected daily in dump trucks, creating a plentiful supply of highly pure iron ore (which, when Gabor was writing, was expected to be in short supply by now unless an alternative mining technique had been developed). This plan never really went anywhere, no doubt because even the most ambitious engineer realized that cracking the earth's crust and unleashing the trillions of tons of hot magma beneath might not be such a good idea.

To get an idea of what this unchecked hubris looks like in practice, consider the effects—and underlying causes— of Hurricane Katrina in 2005. Along with being a tragedy of almost unprecedented scope in U.S. history, the near annihilation of New Orleans by Hurricane Katrina was also a hard lesson in unintended consequences. The Army Corps of Engineers has long maintained a levee system that channels the Mississippi River's flow into the Gulf of Mexico at a high rate of speed, like a garden hose shooting a jet of water. As a result, sediment that would normally be deposited in coastal wetlands is washing out to sea and thousands of square miles of Louisiana would-be wetlands are being lost. Historically, those wetlands served as a buffer for the Crescent City, forming a natural seawall against powerful storms like Katrina.

The levee program began decades ago, when engineers and planners had a lot more ambition and a lot less humility when it came to tinkering with landscapes that had formed over tens of thousands of years. Today, at least in the industrialized democracies, you can't even dredge a river without conducting years of studies to address the concerns of environmentalists and other "stakeholders." Had those same environmentalists been around a generation ago, they would have had a field day with some of the proposals on the drawing boards of the era.

Prediction: Fusion—as in Electricity, Not Cuisine

Fission works by splitting apart uranium, the heaviest natural element. Fusion works by joining together atoms of hydrogen, which is the lightest element on the periodic table (as anyone who managed to stay awake during high school chemistry may recall). Fission can produce a lot of energy, whether under controlled circumstances—as in a power plant—or in a bomb. Fusion, by contrast, can produce a heck of a lot more energy. The fission bomb of the type dropped on Japan is like a firecracker in comparison to the hydrogen bomb, which is like a stick of dynamite. (Thankfully, though, the H-bomb has never been used against human beings.)

There's another significant distinction between these two processes. The first power plants using fission were developed within a decade after the first A-bombs were built. But it has been nearly sixty years since man first detonated the most powerful weapon ever, and we're still not even close to building a fusion power plant. And most scientists today believe we won't get there until about 2050—around one hundred years after the first man-made fusion reaction.

That is a shame, because fusion really is the perfect energy source. Hydrogen, fusion's fuel, is the most abundant material in the universe, so no nation could ever corner the supply— there would be no OPEC-style cartels in a fusion-powered world. Fusion, a U.S. Atomic Energy Commission scientist said in 1968, "could supply the earth with enough power at the current rate of consumption for 20 million years." It's also as clean as solar or any other renewable energy source; unlike its cousin, fission, there would be no radioactive waste. The only by-product of controlled fusion is helium—the stuff you find in toy balloons.

Fusion occurs in nature within the cores of the sun and all the other stars. These celestial bodies formed from small globes of gas and dust that gradually grew larger and larger thanks to gravitational attraction. As these balls of gas got bigger, the pressure at their centers increased until atoms of hydrogen were forced together, releasing heat that we see in the form of starlight and sunlight. That, literally, is how a star is born.

There was tremendous enthusiasm about this energy

source in the decades after the first hydrogen bomb, "Ivy Mike," was detonated in 1952. At the optimistically named International Conference on the Peaceful Uses of Atomic Energy, held in 1955 in Geneva, the scientist leading up the symposium, Homi J. Bhabha, predicted that energy would be obtained from the fusion of light elements by 1975. This possibility was taken seriously outside the rarefied world of academia, too. Speculators at the time were bidding up the price of lithium, a light element that would also be suitable for fusion. Writing in *Scientific American* in 1966, another scientist, R. F. Post, said that there is "little doubt" that fusion "would one day be a reality." Isaac Asimov, writing on the twenty-fifth anniversary of the Hiroshima and Nagasaki bombings, "suggested" that fusion reactors would be built by about 1992.

And a 1964 RAND Corporation study, consisting of a poll of leading thinkers, stated that there was a 75 percent chance that there would be fusion plants by 2010. RAND's methodology, as it turns out, was fairly sound, as this same group was correct about a number of other developments. For instance, most respondents also believed there was a high probability that personality-altering drugs would be in common use by now, a far-out prediction at a time when the advent of the Prozac nation was still decades away.

The big technical problem that has to be overcome, at least in traditional fusion prototypes, is generating heat of a few hundred million degrees, the temperature at which fusion of the lighter elements occur. You also need a way to prevent that heat from incinerating everything within a few hundred miles. In theory, the best way to do this is with a magnetic "bottle," a space contained by a powerful magnetic field in which the fusion reaction would occur. In practice, however, building a useful one has proved to be a very tough nut to crack.

The problem is that the containment systems that have been devised need a lot more power than the reactions they're producing. Controlled fusion has been achieved—at one point leading scientists to believe they were close to a break-through—but the systems developed so far don't deliver a net surplus of energy. It would be possible to make the reactions more powerful, of course, but no one would be left alive to pop the Champagne corks.

There may yet still be a way around this problem. As far back as the 1920s a few scientists were fiddling with "cold," or room-temperature, fusion, a chemical process in which hydrogen atoms are joined without the heat that the reaction requires in nature. Two researchers, Martin Fleischmann and Stanley Pons, created a huge stir in 1989 when they announced that they had achieved cold fusion in a device not that different from a car battery. Other scientists, however, quickly killed the party. No other researchers were able to replicate the duo's results, and today, although the idea has not been completely dismissed, most fusion research is where it has been since the 1950s—focused on ways to safely re-create something like the center of the sun here on earth.

Prediction: Today's Forecast: Whatever You Want

When Bernard Vonnegut discovered a highly effective cloud-seeding technique in the 1940s, he helped launch the science of weather control. He may also have inadvertently served as muse for his far more famous kid brother, Kurt, who would go on to write such classics as *Slaughterhouse Five*, *Cat's Cradle*, and *Breakfast of Champions*.

Kurt Vonnegut frequently wrote about the downside of technology. In *Cat's Cradle*, for instance, military planners manage to invent a superweapon so potent it wipes out everyone—friend and foe alike. In *Breakfast of Champions*, one of the characters has to wade through a stream filled with a toxic goop that leaves his legs encased in plastic. Vonnegut brought home the point that we could be too smart for our own good— a point nicely illustrated by the promise of weather control.

Starting immediately after the elder Vonnegut's discovery, a slew of experts over the next several decades were predicting that we would one day be able to make it rain and snow on demand. We would be able to sap hurricanes and tornadoes of their strength. Presiding over all would be a government "weather authority" that would decide who gets rained on, and when.

A few even speculated that humans could one day alter not just the weather—which is a brief, localized phenomenon— but the very climate itself. We would be able to short-circuit the Indian monsoon season. We could make Scandinavia a tropical paradise. Or, scientists later realized, we could screw up the climate so badly that we could render entire swathes of the planet uninhabitable.

In a 1954 article titled "Tomorrow's Tailor-Made Weather," H. T. Orville, a U.S. Navy meteorologist who served as chairman of a presidential committee on weather control, wrote that by the mid-1990s scientists might be able to plot out the day's weather just as they would a train schedule. Rain would be dispelled so baseball games and Sunday picnics would go on as planned. Farmers would no longer have to fall back on that time-honored remedy, prayer, to provide for their parched crops.

The breakthrough that excited Orville and many others at the time was the discovery that seeding clouds—also known as "cloud-busting"—could cause precipitation to form. Silver iodide crystals dropped from planes or shot from the ground using cannons attract tiny droplets of moisture until the budding raindrop becomes so heavy it falls to the earth. By the mid-1950s, cities and regions that had experienced drought, such as New York City, Dallas, and parts of Oklahoma, were hiring cloud busters to bring their residents relief. The U.S. government began spending millions on weather-control research. A decade later, the National Academy of Sciences, after a two-year study, concluded that seeding could potentially increase precipitation by 10 to 20 percent.

Cloud seeding was believed to be a highly versatile technique, capable not just of causing rain but of drying out clouds, too. Scientists speculated that "overdosing" clouds with silver iodide particles would cause lots of tiny raindrops to form, each too small to fall from the sky. Officials who worked on weather control, pointing to figures showing that flooding caused hundreds of millions of dollars in damage every year, believed that we would one day stop the rains long before the local rivers crested.

This concept could also be employed, it was thought, to weaken hurricanes before they made landfall. Hurricanes are complex weather systems that require just the precise mix of temperature, moisture, and atmospheric pressure to form. Dropping the proper amount of rainmaker crystals into this blender of air and moisture, many scientists believed, would throw the whole system out of whack, causing destructive storms to dissipate. In the early 1960s, this possibility was explored under the aegis of Project Stormfury, a U.S. government program (they really knew how to name secret government projects back then). Pilots dumped crystals into at least two hurricanes, causing a brief alteration to the "eyewall," the layer surrounding a hurricane's heart. Although no one stopped a storm in its tracks, researchers were initially encouraged by the results.

A related effort with an equally fearsome name, Project Skyfire, held out the promise of one day short-circuiting lightning. Scientists tried to induce cloud formation near thunderstorms, drawing away some of the storm's power and thereby weakening the whole system, an effort that had only mixed results. Using a similar principle, scientists believed that they could one day neuter tornadoes as well.

Ocean currents are one of the most powerful engines of climate on the planet because they circulate warm air and water northward from the equator. Ocean currents are the reason why Great Britain doesn't resemble Alaska, even though it's as far north. They are essentially rivers of water within the sea. A few engineers in the 1960s believed we could build seafloor dams to redirect currents wherever we wanted, making historically cold places warm and chilling balmy regions.

Employing a variation of this technique, Roger Revelle of the Scripps Institution of Oceanography envisioned a time, perhaps by the early 1980s, when heated water from nuclear power plants would be pumped into the Pacific off the Northern California coast, forever banishing San Francisco's trademark fog and making the beaches of Marin County as inviting as those farther south in Los Angeles.

Civilians weren't the only ones interested in the possibilities of weather manipulation. After all, one typical thunderstorm releases enough energy to power all of America's appliances for a year. A hurricane packs a wallop equivalent to hundreds of nuclear bombs. The generals and admirals of the 1960s began including these potential superweapons in their arsenals—and even once used weather control in a war being fought at the time.

Under an operation code-named Project Popeye, the U.S. military in the mid-1960s seeded clouds as the annual rainy season was drawing to a close in Vietnam, extending the downpours by several weeks. The idea was to render the network of hidden jungle paths known as the Ho Chi Minh Trail impassable, preventing the easy movement of troops and supplies from communist North Vietnam into South Vietnam.

But this and most of the other, more ambitious weather manipulation schemes gradually petered out. As scientists were learning more about affecting the weather, they were also beginning to grasp just how much weather is a drama with a lot of players; changing one could produce unforeseeable, and horrible, results. Project Popeye was terminated because no one was sure if there would be a long-lasting effect on the rainy season in Southeast Asia, potentially harming not just the enemies in North Vietnam but America's allies in the south, too. The United States and other nations shortly after signed a treaty banning the use of weather manipulation as a weapon of war. (And not too soon, either: some were thinking of deliberately creating tsunamis and tidal waves, using nuclear bombs set off on the seabed, against a coastal enemy.)

Scientists also stopped seeding hurricanes, out of fear that the hurricane's power would just be redirected somewhere else. Botching an attempt to tame lightning could result, ironically, in an even more powerful lightning storm. Deliberately messing with the earth's ocean currents could have consequences worthy of a Kurt Vonnegut novel.

Moreover, weather control methods such as cloud-busting simply proved to be a lot less effective than originally advertised. Scientists gradually learned that it would actually be pretty difficult, if not impossible, to make or stop rain where the conditions aren't already just right (which is why most regions of the world are still at the mercy of droughts or torrential downpours). There has been considerable success in making snow over mountain ranges in the American West, while seeding is routinely used at airports to make hailstones smaller. Special heaters and other techniques are also used at airport runways to clear fog (something first perfected at Royal Air Force bases in England during World War II). All of this, however, is a far cry from actual weather "control." Although research continues, today what the best man can do is give Mother Nature a nudge in a particular direction, increasing the likelihood of rain or snow but by no means guaranteeing a specific result.

As it turned out, the only climate change we've been able to pull off is the kind nobody wants—global warming from carbon dioxide emissions. When scientists first dreamed of affecting the weather fifty years ago, though, that's not quite what they had in mind.

Prediction: We Talk to the Animals

Biologists spend years observing animals, trying to discern aspects of their behavior. But what if there were a more straightforward way of finding out what's on the mind of a bonobo or killer whale—such as asking them? By now, a few scientists believed, we would all be Dr. Doolittles, able to talk to the more intelligent animals.

One of the most prominent proponents of this view was John Lilly, a talented, erratic scientist who has a number of patents to his name but was also one of the first advocates of LSD use. Lilly also perfected the isolation tank, where subjects could supposedly commune with their inner selves and learn secrets imprinted in the very cells of their bodies. Lilly, who believed that human beings had untapped inner potential, was written off by some as an acid-dropping eccentric; to others he was a visionary. Not surprisingly, he was also at the forefront of research into interspecies communication.

In the early 1960s, before he became better known as a New Age guru and sherpa of the human subconscious, Lilly had built a laboratory in the Caribbean where he tried to plumb the depths of the dolphin mind. Although Lilly wasn't the first researcher to note it, dolphins' brains are as large and seemingly as complex as man's, suggesting that their behavior could be every bit as nuanced as a human being's. Lilly, who died in 2001, was among the first to argue that dolphins and whales belong to sophisticated underwater civilizations.

Dolphins are indeed incredible animals. By the time Lilly was conducting his research, the U.S. Navy had trained dolphins and sea lions to perform a stunning range of tasks, such as locating underwater mines and assisting in rescues. The Soviet Union had taken this a step further, making Flipper into a kamikaze pilot. The Russians had trained cetaceans to engage in attacks against enemy ships and subs, the ultimate stealth weapon since an approaching dolphin (which would have a powerful bomb strapped to its back) would not likely arouse suspicion. Dolphins, commonly perceived as friendly and protective of humans, were even taught to attack enemy divers.

Most mainstream animal researchers attribute these feats to simple training, however. As humanlike as rescuing a diver may be, dolphins are really doing something not that much different from a dog who has been taught to sit and fetch. An animal can learn to associate a given command with a given action, but this is not quite the same thing as teaching a person particle physics or European history. The big difference is that people, unlike even the smartest animals, can think abstractly. Humans, for example, can plan for the future, which, by definition, requires the ability to imagine a set of circumstances that hasn't occurred yet. Any behavior that might seem to indicate that animals can envision tomorrow— for example, a squirrel gathering nuts for the winter—is considered by most researchers to be nothing more than evidence of instinct, not forethought.

Lilly was convinced otherwise but knew that it would take a dramatic example to change most everyone else's thinking. The best way to do this was to let the porpoises and whales speak for themselves.

At Lilly's lab in St. Thomas, Virgin Islands, his assistant, a young woman named Margaret Howe, spent two months in the early '60s in a specially designed pool with "Peter" and a few other bottlenose dolphins, attempting to initiate true communication. Lilly—who specifically chose an attractive woman to work with the male dolphins, on the theory that a member of the opposite sex would have more success reaching out to the animals—believed that the "clicking" sounds dolphins make are a rudimentary form of language. Howe spent weeks trying to translate the wails and whistles of "dolphinese" into English. Howe and Lilly also believed that one dolphin had the ability to teach others to communicate with their two-legged, "dry living" instructors. Lilly predicted that, if research were to continue, there would be regular communication between people and porpoises by the early 1970s, a forecast he repeated in the early 1980s and at other times right up to his death.

Lilly believed that a big problem to be overcome was not that the animals weren't smart enough to talk to us, but that we were too dumb—or too uncivilized. Lilly thought that sperm whales, for instance, were so bright that we would bore them. Lilly also thought that these gentle giants avoided us because of our tendency to make war, that we were too savage—animalistic, if you will—for meaningful conversation. When asked why, if they're so smart, dolphins and whales aren't ruling the world, Lilly responded that the animals were "too wise" to do so.

Even more obvious candidates for interspecies communication are chimpanzees and other primates, since, genetically, they're much more similar to people than porpoises are. Taking a page from Lilly's playbook, researchers throughout the 1960s and '70s "adopted" chimpanzees, taking them home and raising them like human infants in the hope that the developing chimps would absorb language naturally, just like babies do. In 1966, Vera Gatch, a scientist at the University of Oklahoma, began to care for "Mae" and "Lucy," two chimpanzees who were deliberately sheltered from other members of their species (so the two would never acquire "bad" chimp habits). The chimps were toilet-trained, wore human clothes, sat at a table for meals, and received language instruction.

Researchers have claimed that primates raised under these conditions eventually mastered a vocabulary of up to two hundred words, enough so that the animals can carry on simple conversations. Others are skeptical, however, noting that teaching a monkey to point to a picture of a banana when it is hungry is hardly evidence of sophisticated communication.

Still, the results were sufficiently encouraging to lead to predictions that one day, apes might actually be competing for jobs with us less hairy primates. Sir George Thomson, in his book *The Foreseeable Future*, believed back in the late 1950s that monkeys would be trained to work picking crops and toiling on assembly lines. Others foresaw strong, agile chimps becoming miners and domestic servants (an idea that no doubt inspired the *Planet of the Apes* books and movies). In 1967 one group of forecasters envisioned apes being trained to drive cars, to benefit "housewives" who would prefer living servants to robots. The forecasters believed the simian drivers would be so well trained that there would be fewer car accidents. Not only could chimps peel bananas with their feet, they would also make better drivers.

Another grisly possibility widely discussed since the 1960s, right up to the present day, was the creation of "chimeras," animals that have been genetically endowed with humanlike intelligence. In this nightmare scenario, cheap labor would be manufactured in laboratories; as in H. G. Wells's novel *The Island of Dr. Moreau*, these creatures would exist in a twilight between the human and animal kingdoms, although they would be slaved to human beings. Today scientists routinely create chimeras—giving, say, mice a humanlike immune system for research purposes—but there have been no reported attempts to put anything like a person's smarts in a gorilla's body.

Chapter Six

HOME SWEET HOME

In the 1927 film *Metropolis*, the workers live in vast underground warrens, while the elites occupy the upper floors of luxurious, Art Deco skyscrapers. In the late 1960s cartoon *The Jetsons*, everybody, even the working stiffs, lives in apartment blocks perched atop massive towers. In the Los Angeles of *Blade Runner*, an early 1980s film, something resembling the pyramids has made a comeback. Yet another recurring architectural style in popular science fiction is the dome, which encloses the city and protects it from unbreathable air or, maybe, surly mutants unable to scrape together a down payment on a nice condo.

For decades, science fiction has primed us to expect cities to become radically different. In fact, though, the last great innovation in city living has been the skyscraper, which was developed more than a century ago. Although building techniques are more advanced, a time traveler from 1950 would probably find modern Chicago or London fairly recognizable. A lot may have changed, but the skyline has pretty much remained the same for decades. The city was expected to change in other ways: depending on whom you asked, it was either going to become much larger, fade into irrelevance, or be removed from the surface of the earth entirely.

To accommodate the billions more people who would be populating the planet by now, some thinkers felt, would require a radical rethink of the very idea of community. The city itself, for one thing, would grow to the point where it would become the "megalopolis," a unified home to tens of millions that would span state and even national lines. In this world, places such as Chicago or Los Angeles would be mere pearls on a string of linked urban centers, entities that would require new names like "Los Sandieglos," "San-San" (a reference to San Francisco and San Diego), or "Boswash" (Boston–Washington, D.C.).

With millions flooding into urban cores, the city was expected to grow up as well as out—New York was going to be home to tower blocks housing tens of thousands of people, vertical villages that contained schools, shopping and houses of worship. Lots of people by now were to be spending virtually their entire lives under one roof.

And where was it written that the city even had to be aboveground? Thanks to advances in submarine technology, and given the mineral wealth to be found under the sea, tens of thousands of us were to be living below the ocean waves, operating mining colonies and catering to terrestrial tourists. Caves, albeit of a modern, 21st-century variety, would also make a comeback, as millions took advantage of the fact that an underground dwelling can easily be kept cool in summer and warm in winter.

In fact, some thought, why not make conventional cities climate-controlled, freeing them forever from the insults of Mother Nature (or air pollution)? In the 1960s there were plans to put entire sections of Manhattan beneath a dome, enabling city planners to maintain a nice, balmy 72 degrees year round. That is, of course, assuming the city survived in its current form—others believed that the world's urban areas would depopulate, supplanted by slightly more urbanized suburbs known as "New Towns," a few examples of which have actually been built.

The individual home itself wasn't immune to speculation, either. Obsessed with making the life of the "housewife" less dreary, manufacturers back in the '50s and '60s envisioned houses that did the cooking, cleaning and even gave their occupants a checkup every now and then. Dishes and even furniture wouldn't be cleaned anymore, either—just thrown out and replaced (this was an era when there was a vogue for cheap, throwaway plastic and paper goods).

And what would that house be made of? Seemingly anything but good old-fashioned brick and wood. Instead, some thinkers thought our castles would be built of steel, or glass. Prefabrication would enable anyone who could hold a hammer to slap together their domicile in an afternoon. You could even go with a chalet of plastic, an option that would set you back just a few thousand dollars.

The megacity, however, was where most thinkers believed this home would be located. Today many of us were to be residents of gigantic city-states hundreds of miles in length containing tens of millions of people.

Prediction: Megalopolis—Welcome to Boswash and San-San

Back in 1898, the Big Apple got bigger when the formerly independent city of Brooklyn was downgraded and became a mere borough of the city of New York. With cities getting larger all the time, forecasters back then figured that every metropolis would follow suit, until one day we would live in communities more populous than most American states—or even some countries. Someone living in upstate New York would belong to the same "hometown" as someone living in eastern Maryland. As it turned out, though, the big city only got so big before residents started heading for the hills (or the suburbs).

In 1903, William Baldwin, president of the Long Island Rail Road, predicted that "Philadelphia would be a suburb of New York City in 20 years." In his seminal work of futurism, *Anticipations*, H. G. Wells believed that in the 21st century the "great stretch of country" between Albany, New York and Washington, D.C., would come to be considered part of the metropolitan regions of New York City and Philadelphia. London's metropolitan area, meanwhile, would encompass nearly all of southern England and Wales, covering a territory almost equal in area to Ireland.

The concept Wells described wasn't given a name until the early 1960s: *megalopolis*, a string of connected, highly populated urban areas, such as the eastern seaboard of the United States. At the time, this section of the country was virtually unique in the world.

But today, as most demographers predicted, there are dozens of megalopolises on the planet—indeed, virtually all of South Korea is considered one. The rise of the megalopolis has occurred, as you would expect, in lockstep with urbanization in general—by the early 21st century, for the first time ever, more than half the world's population lived in urban areas, a figure that is expected to grow.

Where many demographers erred was in overestimating the size of these regions and how tightly linked their constituent parts would be. One expert, David Lewis, believed in the mid-1960s that mass transit of various kinds—whether air or rail—would get commuters back and forth to work at 1,000 miles per hour, a speed that would make it possible for a person in southern New Hampshire to get to the office in Washington, D.C., with time to spare. Others believed that, in the advanced countries, at least, the cities that make up the megalopolises would be much more attractive than they proved to be. A population projection in the late 1960s, for instance, showed the New York City–northern New-Jersey–Connecticut region rising to 30 million people—a guess that overshot the mark by about 10 million. Other cities in the so-called "Boswash" corridor, meanwhile, actually lost population in the last few decades of the 20th century. The futurist Herman Kahn believed in the late 1960s that this area would be home to something like 80 million people, but it actually contains just shy of 60 million today.

Other megalopolises came up short, too, such as the one stretching through the Rust Belt ringing the Great Lakes, which includes struggling cities like Buffalo and Cleveland. Most demographers at the time did not foresee the depopulation these cities would experience as the traditional smoke-stack-industry jobs that sustained them evaporated. Instead, Kahn believed that millions of people in the Great Lakes region between Chicago and Pittsburgh would live in an entity that might be called "Chi-Pitts"—or maybe something that had a nicer ring to it—while in California, the mega-hometown of the future would be known as Los Sandieglos, or maybe "San-San" (as in Francisco and Diego).

Another factor that has hampered the growth of the megalopolis has been the rise of the "exurb," low-density towns that have sprung up in the last few decades that are located fifty miles or more from the major job centers. These communities have been built in what had been farmland around Washington, New York City, Atlanta, Los Angeles, and elsewhere, populated largely by families that want homes and yards but can't afford to buy close to urban centers or the inner suburbs. The construction of these towns has literally been fueled by cheap gasoline, as some commuters think nothing of driving 150 miles a day to get back and forth to work.

Another development that few foresaw decades ago was the huge rise in cities outside the traditional megalopolises, cities such as Charlotte, North Carolina, Salt Lake City, and Las Vegas. For most of America's history people moved to the manufacturing centers, going where the jobs were. But with the rise of the service economy the jobs, to an extent, began to follow the people, as companies set up shop in cities with a low cost of living precisely because there would be both a ready-made workforce as well as lots of customers. America's megacities, such as New York and Los Angeles, are still growing, but these cities are also losing longtime residents who are moving to greener, less expensive pastures well outside the nation's massive population corridors. We haven't quite made the transition from plain old metropolis to megalopolis just yet.

Prediction: Home Is Under the Dome

In a future where the air in cities was going to be unbreathable, some forecasters believed we'd come up with a pretty ingenious solution. Like cleaning up the pollution, you ask? Not quite.

According to some experts back in the 1960s, many of us today were supposed to be living in what would have looked like the world's largest snow globes: domed, climate-controlled, and pollution-free cities.

If "the future" once had a signature geometric shape, it was probably the geodesic dome. Popularized by the futurist and inventor R. Buckminster Fuller, these are the structures that can be seen today atop modern sports arenas or at science museums. Consisting of struts arranged in geometric patterns across a sphere, geodesic domes—perhaps the most famous example of which can be found today at Epcot Center in Florida—represented a revolutionary advance in construction. Lightweight but strong and relatively easy to assemble, Fuller and others believed that the domes could be built to massive proportions, one day housing entire colonies of people on the moon and Mars.

And what would work on other planets would work equally well here on earth, many also believed. Thousands of people who attended the 1967 World's Fair in Montreal, known as "Expo '67," saw a massive, twenty-story geodesic sphere designed by Fuller at the American Pavilion. Such domes, Fuller said at the time, would let in sunlight and moonlight, but "the unpleasant effects of dust, heat, bugs, glare etc. will be modulated by the skin to provide a 'Garden of Eden' interior." One enthusiast of Fuller's concept, an executive with the Canadian Refrigeration and Air Conditioning Association, named T. W. McClorg, envisioned a time when Toronto would be under glass, enjoying Caribbean-style weather year-round. "I suggest we can look forward to the day when a born-and-bred

Torontonian will be able to grow oranges in his back garden, watch his roses bloom in December . . . buy a top-less convertible for his wife, laugh when he gets his heating bill and hang up his snow shovel forever," he wrote in 1967.

Another fan, Robert Ingersoll, the chairman of Borg-Warner Corporation, predicted around the same time that by the early 21st century a number of cities would be tucked under domes, and those communities would be completely climate-controlled; people would enjoy springtime temperatures without allergies because pollen would be filtered out. Dome-ites would be healthier in general, too, thanks to filters that would screen out disease-causing bacteria.

Skeptics, of course, abounded, especially when Fuller in 1965 proposed enclosing much of Manhattan Island beneath a semisphere. But Fuller—a polymath and one of the most creative American thinkers of the 20th century—had anticipated many of their arguments. When it was pointed out that a New York City dome would cost hundreds of millions of dollars, Fuller noted that the city spent that much in snow removal in a decade; a dome, therefore, would pay for itself in ten years. Fuller also pointed out that much of the heat used to warm the city's buildings is wasted, "leaking" through roofs and into the atmosphere. A dome would capture that heat. And domes benefited from economies of scale—the bigger you made them, the cheaper it was, on a per capita basis, to provide heating and cooling for residents. The one Fuller envisioned for New York would have been a mile high and two miles in diameter, soaring above the Empire State Building and enclosing something like a million people.

Although widely recognized as a brilliant, visionary idea, Fuller's concept of a domed city never really caught on. For one thing, engineers discovered that domes of that size would actually be much more difficult to maintain than Fuller originally believed. And in any event, providing Canadians with year-round springlike temperatures or freeing New York's public works department from snowplowing weren't enough of an incentive to place entire cities under glass.

Prediction: Cities Underground, Underwater

Placing a dome over a city was considered not only a good idea for those communities that already existed on land, but also a way to create communities in entirely new environments: like belowground or on the ocean floor.

By the 1950s scientists had learned that the vast deeps contain billions in mineral wealth; even today, a fair chunk of the world's oil comes from seaborne drilling platforms that resemble, if not towns, then small villages. Much like San Francisco and Seattle—cities that grew from Gold Rush–era boomtowns—several forecasters thought that by now, thousands of modern-day prospectors would be toiling deep beneath the waves, reaping fortunes from magnesium, zinc, and other valuable metals. The Wild Depths would be the successor to the Wild West.

A lot of factors back in the 1960s fueled interest in creating a modern Atlantis. Treaties brokered by the United Nations regarding the ownership of the ocean floor helped define national and "free" areas of the seabed; according to some estimates, these treaties meant that about 80 percent of the ocean floor lay outside any given national boundary, so these areas would belong to whoever could exploit them. Submarine and underwater construction technology had vastly improved. The aerospace industry, meanwhile, recognized that many of the same technologies that could be used in space could also be employed in the equally alien atmosphere of the deep ocean.

By the early 21st century, thousands of people were to be living within reinforced domes on the seafloor hundreds of feet beneath the waves. These homes were to be serviced by "aqua-cars," relatively inexpensive submersibles that would supply these communities with whatever they needed from the terrestrial surface—which wasn't expected to be much. The communities, which many believed would be operating undersea oil operations at first, would be growing or catching their own food, while electricity would come from generators powered by the waves. One French architect even designed a self-sufficient community that would be located beneath the River Seine in Paris.

When it came to the oceans, no speculation was seemingly too bold. "Within 50 years," F. N. Spiess of the Scripps Institution of Oceanography said in 1966, "man will move onto and into the sea—occupying it and exploiting it as an integral part of his use of this planet for . . . minerals, food, waste disposal, military and transportation operations and, as populations grow, for actual living space."

A contemporary of Spiess, Athelstan Spilhaus of the University of Minnesota, believed that the U.S. government would create "sea-grant" colleges, counterparts to the land-grant institutions founded in the 19th century to conduct research into agriculture, mining, and other practical subjects that helped settlers exploit the American continent. There would be "county agents" and an "Aquacultural Extension Service" spreading learning among the scattered colonies and submersibles of the 21st-century aqua-nation.

It wasn't to be all work and no play in the oceans, however. General Motors' 1964 Futurama exhibit promised that by now, vacations beneath the sea would be fairly common— probably about as typical as a family from Milwaukee spending a few weeks in Hawaii is today. Consider the following passage, an excerpt from the script for the exhibit: "And in warmer seas are new realms of pleasure, a weekend if you wish at Hotel Atlantis, in the Kingdom of the Sea. A holiday of thrills and of adventure; of beauty and enchantment; of radiant wonders; in the sun-bright gardens of the sea." Picking up on this theme, one prognosticator believed that an undersea Ramada or Motel 6 would be built in Florida by about 1990. (As of this writing, undersea resorts are being planned at a Fijian island and in Dubai, luxury hotels catering to fairly wealthy guests.)

Some even believed that, as underwater living became more popular, science would turn some of us into Aquamen and Aqualadies, people biologically adapted to live under the sea. Jacques Cousteau, the famed French undersea explorer, thought that one day man would be endowed with gills via genetic or surgical manipulation, enabling him to be free of the protective gear required for spells underwater. Scientists in the 1960s tested an artificial skin, resembling Saran Wrap, that would encase divers and enable them to extract oxygen from the surrounding water; the device was tested on a mouse that was able to breathe for half an hour while submerged.

Today, undersea mining operations are commonplace, and explorers in submersibles can go even deeper than was possible a generation ago. But a real deep-seafloor community, where large groups of people spend weeks and months in the briny depths, has yet to be built. Thanks to advances in robotics, it's proved to be a lot cheaper to send unmanned probes instead. Similarly, mining and drilling technology today is advanced enough so that it's really not necessary for human beings to be encamped several hundred feet below sea level.

Although it was not nearly as popular as the underwater variety, a few urban planners were also promoting the idea of underground cities a generation ago. The arguments were all strictly practical: as population increased, land at the surface would become ever more scarce, while sprawl—simply building farther and farther away from population centers—was considered undesirable. Submerged communities could also save a fortune on heating and cooling because they would always be insulated from freezing winters and scorching summers. A *Wall Street Journal* poll conducted in 1967 found that many advocates cited benefits such as low noise levels (no one living underground would be kept awake by passing airliners) and low maintenance costs (no one would have to paint the house).

Today large, underground retail areas can be found in cold-weather cities such as Helsinki and Toronto, while cities such as Beijing boast vast networks of tunnels. Subterranean garages and other structures are also more common than they were a generation ago. But "mole people"—like aquanauts and dome-ites—for the most part don't exist today. Nearly all of us live on solid ground, under open skies, with nary a dome in sight—unless you count the local football stadium.

Prediction: Movin 'Out—the Incredible Shrinking City

"Decentralize. Decentralize. Decentralize. Giant cities are a feudal hangover. They'll vanish and leave grassy streets."

When the groundbreaking architect Frank Lloyd Wright expressed this sentiment in the late 1950s, the American city had yet to be ripped apart by widespread race riots. Plentiful blue-collar jobs still supported a huge, urban working class. The tax base was still large enough to provide considerable investment in roads, bridges, and other infrastructure.

Within a decade, however, many thinkers believed that Wright's admonition was becoming prophecy. No one would have to force city dwellers to leave; they were voluntarily abandoning the cities in droves for the suburbs. Downtown was becoming a ghost town.

In the late 1970s, Theodore White, a prominent journalist and author of *The Making of the President—1960*, was among many who sounded the death knell of the contemporary American city, forecasting that New York would lose almost half its population by the early 21st century, with a similar fate befalling Boston, St. Louis, and every other major metropolitan area. He believed that by 2000, the cities—thanks to a vastly diminished tax base—would have to be quasi-federalized, with the entire country helping pay for urban police forces and hospitals.

A decade earlier, author Stuart Chase also believed that the nation's urban areas had maxed out, predicting, as White would later, that New York's population would drop to below 5 million. In a flight of speculative fancy, Chase, in his book *The Most Probable World*, described a "Great Demolition" occurring in New York in the 1980s, with half of the city's buildings replaced by greenswards and parks. The former New Yorkers had moved to so-called New Towns, rigorously planned communities of about one hundred thousand that fell somewhere between a large suburb and a small city. These New Towns—

a very popular concept among urban planners in the '60s and '70s—were to combine all the benefits of the suburb and the metropolis while having none of their drawbacks. Racially integrated by government decree, New Towns would not suffer from white flight. Population density would be high enough to justify mass transit, reducing the need for the pollution-causing car, but not so high as to be unmanageable. The New Towns would also be built close to existing cities, counteracting the effects of sprawl and ensuring that the communities would still be close enough to benefit from urban, cosmopolitan culture. The model for this concept, at least in America, was Columbia, Maryland, which was completed in 1967. Originally intended to be a cutting-edge example of progressive, 21st-century living, Columbia has actually evolved into a fairly conventional suburb of Baltimore.

Not everyone had written the city off, however, with many predicting what could accurately be described as "hyperurbanization." One commentator, Ferdinand Lumberg, predicted back in the 1960s that by now it would be the suburbs beginning to experience significant decay: "The suburbs themselves, now beset by problems similar to those of the cities . . . will begin to deteriorate physically owing to poor construction, will be taken over by the lower economic classes." The displaced suburban middle class, meanwhile, would be flooding back into the urban core they had fled in the middle decades of the 20th century.

As the urban population expanded, cities would have no room to physically grow—at least not horizontally. Instead, tens of thousands were to be living in huge tower blocks, such as the ones described by futurist Theodore Gordon forty years ago. One giant building would contain schools, hospital, and police and fire departments. "It may be possible to be born, live a full childhood, receive an education, work profitably and finally die in a single environmentally controlled building," he wrote. A British engineer, Wilem Frischmann, even proposed creating an 850-story building (about two miles high) that would be home to half a million people.

These cities-within-cities, linked by climate-controlled walkways, would rise above green plazas and transport would be provided by mass transit (the car having long been banned from the city center).

In the largest cities, renters generally far outnumber owners, leading many to conclude that by now the classic American dream of home ownership would have gone the way of the dodo. With so many Americans living in megacities, there were to have been a lot more landlords by now. Alvin Toffler, in his book *Future Shock*, noted that in 1969, "for the first time ever more building permits were being issued for rental construction than for private homes." (The rate of home ownership in America has actually increased since then.)

Most of these predictions were at least partly correct. Some American cities, such as Detroit and St. Louis, have shrunk drastically—but at the same time, people have flooded to what had been small cities, such as Phoenix and San Jose. The urban center has become attractive again for the upwardly mobile, forcing poorer people into the suburbs—but the suburb is still the main choice for the middle-class family. Growth in America's very largest cities has continued, but not to the point where anyone is planning to build giant apartment buildings of the type seen in films like *Blade Runner*. The city, today, is very much alive and well, but it's neither as huge nor as small as some believed it would be thirty or forty years ago.

Prediction: Move and the House Comes with You

Today a house is the single largest investment the typical family will make. But that's not the way it was supposed to be. Instead our cities were going to be filled with urban nomads, wanderers who packed up the 21st-century equivalent of a tent and followed jobs in sales instead of the caribou herds. Instead of being his or her castle, the modern home was going to be a relatively inexpensive throwaway, something that didn't even outlast the family car.

Houses had to be made on the cheap, it was once thought, because there was a growing demand for them, and millions of people would not be able to afford a wood-framed, three-bedroom number with a backyard. The people who began streaming into the cities from the countryside in the early 20th century were going to have to make do with a lot less.

In his *Anticipations*, for instance, H. G. Wells in the early 1900s thought that a person's home might one day be built in the same way as a homecoming float—with something like chicken wire and papier-mâché: "It ought to be possible to build sound, portable and habitable homes of felted wire-netting and weatherproof paper on a light frame."

Another of the reasons the home needed to be cheap and easy to assemble, as Wells hinted, was to accommodate a population that was increasingly mobile. Throughout the 20th century various experts contended that moving from one fixed, solid address to another would rapidly become outmoded. Instead the middle-class family in Topeka would simply pack up the entire house and take it with them when Dad was hired for a position in Chicago.

In the 1950s, several architects and designers believed that the prefabricated home was the wave of the future; it could be slapped together in an afternoon from mass-produced parts. Nor would people buy a new home each time they moved— instead the parts would come with them, moved to a lot that had once contained a previous owner's prefab home. These homes would consist of simple panels made of plastic or even steel, and adding rooms or floors would not require specially skilled craftsmen. In this vision, if you were able to follow the assembly instructions for a baby's crib, then you would be competent enough to add an upstairs bathroom, too.

Not surprisingly, R. Buckminster Fuller, the futurist who made a number of predictions about future homes, had much to say on this subject. In the mid-1960s he pictured the 21st-century family as wanderers who would no longer be living in fixed, "heavy shells" made of brick and wood. He believed people would actually own five-hundred-pound "life support systems" that provided heating and cooling, which would then be moved to what Fuller described as "tents": "lightweight, movable and mass produced." These "life-support systems," Fuller believed, would cost just a thousand dollars—still quite a bargain even after adjusting for forty years of inflation.

Fuller also promoted his "Dymaxion" house, a concept he developed over the course of fifty years and that was once extraordinarily popular with city planners. Like his domed cities, the Dymaxion house was a prominent feature of many visions of the future abode, partly, no doubt, because it looked just like what a space-age dwelling should—a shiny aluminum globe that resembled the satellites being sent into orbit for the first time. The house was to consist of aluminum panels built around a single, vertical supporting beam from which high-tension cables would be suspended. The dome shape would allow for more efficient temperature control, while water use would be greatly reduced thanks to reclamation systems. The mass-produced Dymaxion house was intended for any climate, from a Minnesota winter to a Texas summer. The concept of course had its critics—including those who were horrified at the thought of entire towns consisting of sterile, metal balls—and, aside from a few model homes, no Dymaxion units were ever built.

A contemporary of Fuller's, an urban planner at the Massachusetts Institute of Technology named Kevin Lynch, believed that in the coming era, which would see the death of the fixed address, people might opt for "symbolic homes," summer vacation cottages or beach houses made the old-fashioned way and that would meet the psychological need for people to have some sense of being rooted to a place. "A symbolic home occasionally visited . . . may offer some stability to mobile people and thereby make change elsewhere more acceptable," he wrote.

Not everyone believed that the stationary home was going to become extinct, but many did think it would look nothing like the houses of the 1950s and '60s. Fifty years ago, representatives of the plastics industry believed that future houses would be built of lightweight foam that would be sprayed into a simple frame until it set—a method that had all the required benefits, including speed, ease, and inexpensiveness of construction. The walls would then be doused with a blast of radiation to give them added strength.

Others, meanwhile, took their cue from the modern skyscraper, designing single-family homes built almost entirely of windows. Like a contemporary take on the classic igloo, some speculated back in the 1950s that we would be living in clear domes made of special, malleable glass. Beneath their transparent bubble, occupants would experience the effect of being outside as they sat in their parlors and at their kitchen tables. For privacy, such homes would likely contain subterranean areas, too, for bedrooms and bathrooms.

Prefab, "factory" homes have become commonplace, especially in rural areas where it's not unusual to see a house being trucked to a development carved out of the woods. But today, when the owner moves, the house stays put, just like in the good old days. New construction techniques have been developed, but houses for the most part are still made of stone and wood. And a few daring souls do live in all-glass houses— but, odds are, not on your block.

Prediction: Sonic Dishwashers, Disposable Sofas, and Other Household Conveniences

You are a "housewife" in the early 21st century, hosting a dinner party. A boozy guest spills his red wine all over your beautiful sofa. Forty years earlier, this would have been a disastrous faux pas. But in the modern age, thanks to a fundamental change in the way we relate to the stuff we own, that drunken clod gets a pass from the hostess and the other guests. The resourceful little lady of the house simply deflates the stained couch and reinflates a gleaming replacement, and everyone joins in a good-natured chuckle at that tipsy friend's expense.

Such scenarios were possible in the world described in a late 1960s TV special, hosted by Walter Cronkite, in which viewers were given a sneak peek into what everyday life would be like for their children and grandchildren. During the special, the beloved journalist conducted interviews with experts who believed that America and the rest of the world was on the cusp of fully embracing a Dixie-cup culture, one where the homey, sturdy heirlooms that people had cherished for generations would be replaced by the disposable everything. Almost nothing in this future home was made to last.

In this world, no one "did" the dishes anymore. Instead plates made of plastic or paper would completely replace conventional flatware. After a meal these plates would be dropped in the sink and would dissolve in water, and you'd go out and buy some more. Or you would scoop up your plates and stick them in a gadget that used sound waves to reduce them to a pliant goo, with the remains of the meal filtered out. The device would then reshape this gunk into plates ready to use for the next meal. You would even be able to set this machine to crank out the "good" tableware for special occasions.

And who was going to do the cooking for that 21st-century Thanksgiving dinner? Not you—your house itself had that grueling task covered. As far back as the 1950s many believed that every appliance in the home was to be connected to an electronic "brain" that did everything from roasting the chicken to peeling the potatoes. Mrs. Future would never get a spot on her apron again as she prepared the family repast—all she had to do was select the appropriate punch card for a given type of meal and then step back as appliances popped out of the counter top and began slicing, dicing, and broiling all on their own.

Other aspects of the traditional household routine were likewise expected to be wholly transformed—sometimes merely by the application of a little common sense. More than one hundred years ago, for instance, Wells believed that rooms would one day be built without corners but with rounded edges, making it a lot easier to dust. Windows would be installed with tiny jets of water so that they would be self-cleaning.

But there were also to be high-tech weapons in the war against grime and grit. In the late 1960s, Raymond C. Sandin of General Electric believed that there would one day be no more need to take out the garbage—beams of searing light would incinerate the trash instead. Electronic dust collectors, others thought, would be installed in every room, or, alternatively, ultraviolet beams would zap stray particles from the air.

Not even the bedchamber was to be immune from this high-tech onslaught. Many believed the bed would be built into the walls or would consist of a plastic box akin, to a baby incubator, that would automatically adjust the temperature and air mixtures during the night for optimum sleeping comfort. Alarms and biofeedback devices would gently rouse you from slumber, and after your feet had hit the floor, the bed would make itself—using disposable paper sheets like those found in hospitals. Toffler quoted a Swedish magazine article stating that the future sleeping cubicle would even have buttons for breakfast or for when the occupant wanted to read.

In this advanced machine for living—*house* doesn't do justice to the speculative 21st-century home—even the humble lightbulb had been relegated to the past. Instead, C. C. Furnas wrote nearly eighty years ago, walls might be coated with gently glowing radium—the stuff that had been used to make watch hands fluorescent and that's also dangerous to human health after long-term exposure. Failing that, Furnas believed, chemists might have learned to manufacture or harvest the chemical that makes fireflies glow and use that for illumination instead. Or, according to a 1955 article by Michael Amrine, your entire home could be swiveled with a touch of a button to catch the most sunlight throughout the

course of the day. (Amrine imagined a husband in 1985 grous-
ing about his wife's lousy "house driving" skills.)

There was considerable skepticism regarding many of
these imagined innovations. Some pointed out that there
might be a limit to how much even Americans would be willing to pay for creature comforts. There was certainly a limit
to how much in tasteful furnishings people would be willing
to sacrifice for convenience's sake; an inflatable sofa was considered tacky by as many people in the late 1960s as it is today.
And it doesn't seem likely that any of us will be painting our
walls with firefly juice or radioactive by-products any time in
the near future

Obviously, some advanced appliances did become popular, such as the microwave oven and the robot vacuum clean-
ers that have been introduced in the last few years. Perhaps
the fully mechanized home has not yet caught on because of
the expense—it seems likely that if someone could develop an
inexpensive, fully automated kitchen, it would probably be a
hit with today's busy families.

Or perhaps, when it comes to the home, there's only so
much modernity we can take in a generation. It could be that
in creating the environment we live in as adults, we subcon-
sciously hark back to the familiarity of childhood, at least to
an extent. After all, it seems unlikely that even the most sophisticated "automatic" kitchen could ever make meat loaf
the way Mom used to make it.

Prediction: The Picturephone, 3-D TV, and the Newspaper by Fax

For a long time, the picturephone—not the cell phone that
takes a picture, but a phone that would allow you to see who

you were talking to in real time—looked like a sure bet, at least on paper. The technology could certainly be developed. What the picturephone's supporters didn't count on, though, was how many of us choose to have our phone conversations in our bathrobes, unshaven, or with curlers in our hair (not to mention how many of us tend to roll our eyes while a long-winded friend prattles on and on). Just because someone wants to talk to you doesn't necessarily mean they want to teleconference.

Back when the elements of picturephone technology were first emerging, however, no one quite appreciated that basic fact. As far back as 1927, AT&T, the American tele-communications company that used to have a domestic monopoly, worked toward the goal of wedding the two devices that would come to be found in virtually every American home: the telephone and the television. Like peanut butter and jelly, the two seemed a natural fit, enabling people to communicate with anyone anywhere while simultaneously allowing them to view whom they were speaking to on a television screen. It was as close to face-to-face, long-distance communication as technology back then could conceivably provide, and Ma Bell worked hard to try to bring that vision to life.

In 1970, Bell Laboratories, AT&T's storied research arm, set up a simple, and very limited, picturephone system among New York, Washington, and Chicago. It was hardly conve-nient: a call between New York and Washington cost sixteen dollars, you had to plan the call in advance with the other party and the Bell operator, and you had to go to a special booth, staffed by a helper, to make the call (New York's was located in Grand Central Terminal).

Still, Bell believed that this was just the first step toward a time when the picturephone would be as common as its two constituent components, and many sober thinkers accepted that vision as virtually a foregone conclusion. A picturephone debuted at the 1964 World's Fair and reappeared at the 1967 Expo in Montreal. Western Electric—AT&T's manufacturing arm—repeatedly ran ads throughout the '60s and early '70s showing picturephone systems, including one featuring a pretty young woman on the screen and the promise that "Someday You'll Be A Star."

And of course sci-fi writers ran with the concept: the picturephone made an appearance in Fritz Lang's classic 1927 film *Metropolis*, comic-book detective Dick Tracy has his two-way wrist TV, and in the late 1970s TV show *Space: 1999*, characters routinely chatted while viewing.

The concept, however, never found its legs. Even in the mid-1970s, by which time telecommunication technology had vastly improved, the service still cost about ninety dollars a month—a fortune in those days. One of the problems with the picturephone involved the dynamics of supply, demand, and investment. Since at least the 1960s, AT&T could have offered picturephones a lot more cheaply—if the demand materialized. But without the demand, AT&T could not make the investment to its communications system to bring costs down—a classic catch-22.

Even when data transmission technology became vastly more advanced and a lot less expensive, however, no one was clamoring for the picturephone in the same way that they would later seek out broadband Internet access or the cell phone. Bell's own research probably provides a clue as to why: back in the mid-1950s, John Karlin, who studied consumer psychology for Bell Labs, found little evidence that people always wanted to actually see each other when they were speaking (although Karlin's research did reveal that they would love push buttons, lighter receivers, pastel colors, and other innovations that would eventually become commonplace). While it's true that today people have all kinds of ways to send live voice and visual—over the Internet or on their cells—it seems most people also want to retain the ability to make their calls without being seen. A picturephone in every home would have made that impossible.

Another long-promised advance would likely be greeted with open arms, however: 3-D television, which has been in the works for at least fifty years. Three-dimensional viewing already proved a hit among 1950s-era moviegoers, who wore cardboard 3-D glasses during horror and action films (a fad that nearly died out completely but has made the odd, gimmicky return in recent years). Televisions have gotten larger, flatter, and crisper, while video game players crave an experience that's ever more realistic. A broadcast image that envelops the viewer would clearly be a much bigger hit than the picturephone proved to be.

Three-dimensional video images were traditionally produced using the stereoscope, a Victorian-era technology in which two images, placed side by side, were shown, creating a composite picture that had depth and seemed to pop off the flat viewing surface. Those cheesy cardboard glasses were later used to amplify the effect when images were shown in movie theaters.

Engineers and scientists have seemingly been on the brink of a breakthrough for years. Back in the early 1960s, *Look* magazine proclaimed that by 1987, television viewers would be enjoying an experience that "will feel as if [the audience] were in the center of the action." The futurist Theodore Gordon speculated that viewers would be wearing a helmet that not only provided a vibrant viewing experience, but an olfactory one as well—this device would also transmit aromas during programs, "to obtain the viewer's complete involvement."

Several years later, in the early 1980s, Japanese electronics giant Sony had developed a prototype for a 3-D system involving special lenses that would actually transmit images directly into each of the viewer's eyes, a technology that one Sony researcher believed would be commercially available by 1990. Around that time another researcher, James Butterfield, president of 3-D Video Corporation, said that 3-D viewing would very soon be a reality; "[By the mid-1990s] you won't know whether the Johnny Carson in your living room is a video image or flesh and blood."

The former *Tonight Show* host passed away long before viewers ever had the chance to literally welcome him into their living rooms via their futuristic television sets, but there's still a remote hope that Carson's successor, Jay Leno, will be popping into American homes. In July 2008, researchers at the University of Arizona announced a breakthrough in the transmission of holographic images, a 1960s-era technology that is in common use on such everyday items as credit cards and driver's licenses. Hologram transmission made a fairly bizarre debut during coverage of the 2008 U.S. presidential election, when hosts at CNN conversed with the video simulacra of the musician Will.i.am, who was "beamed" into the cable TV network's studio.

Rounding out the shoulda-been but never-was telecommunications trifecta is the newspaper by fax. In 1964, long before the technology became commonplace, the RAND Corporation predicted that newspapers and magazines would by 2005 be directly transmitted into the home. With the advent of television many commentators predicted that print journalism would have to embrace new technology in order to stay relevant; the faint outlines of the twenty-four-hour news cycle were already emerging and few thought that newspapers that were delivered the morning after the news broke would survive unchanged.

At the same time, telecommunications companies were offering more and more bandwidth over telephone lines for a lot less money. News organizations had long been getting transmissions of print stories via electronic means such as the teletype. Businesses were clamoring for a way to send contracts and other text instantaneously. The ground was set for the advent of the newspaper by fax.

Timing, however, is everything. Fax machines were commonplace in business offices by the late 1980s and were becoming increasingly popular in the home, too. But before news organizations and publishers had the chance to set up a fax-based distribution network, another game-changing technology emerged on the scene: the Internet. By the mid-1990s newspapers had leapfrogged over fax technology and began publishing online, a vastly more convenient and efficient way to get their product to millions of readers. The newspaper-by-fax concept had become a dinosaur almost as soon as it had emerged from the cradle. And today, the very idea of the printed newspaper is slowly fading out.

Chapter Seven

LIVING, LOVING, EARNING, AND LEARNING

The industrialized, capitalistic West was seething in the year 1888. The economies of America and other countries were rapidly expanding, creating wealth on a scale never seen before. But that wealth creation came at a cost: immigrants to the United States from overseas and migrants from the countryside swarmed into the hot, dank mills and factories, where they worked six- and seven-day weeks for wages that barely kept them fed and clothed. The situation was no better for the average person working the soil, either: many "farmers" had become sharecroppers, scratching in the fields for a pittance that was never enough to retire their debts to the huge landowners (who often paid in scrip, not cash, a form of currency that could only be redeemed at company stores also owned by landowners).

It was an era of breathtaking inequality. Plutocrats amassed vast fortunes nearly equivalent to the gross domestic product of small nations while thousands of female factory workers "were forced to sell their virtue for bread," as one activist at the time put it. In his *A People's History of the United States*, Howard Zinn writes that there were 1,400 strikes in America that year—many put down violently—involving a half-million workers. A growing chorus of labor leaders was calling for the overthrow of the capitalist system in favor of socialism.

One of the witnesses to this seismic turn of events was a successful Massachusetts author and journalist named Edward Bellamy. Self-effacing to a fault yet burning with a need to take some role in combating the injustices around him, in 1888 Bellamy published his best-known work, *Looking Backward*. Its hero falls asleep and wakes up Rip Van Winkle style, 112 years later, to an America where private industry has been completely commandeered by the state, everyone is well educated, and income disparities no longer exist. The place is a paradise.

Bellamy did not believe his book was a Utopian fantasy but rather a prophecy. As Bellamy saw it, capitalism was just a phase in man's inexorable march toward a better world, a world that would come into being naturally and—here was a key point—without revolutionary bloodshed. This optimistic vision gripped the imaginations of millions turned off by the prospect of a violent, Russian-style revolution but equally unhappy with the status quo. Bellamy societies formed across the globe. Among his biggest fans was Eleanor Roosevelt.

Bellamy's ghost loomed over progressive developments such as the introduction of the forty-hour work week, high taxes on corporate income, and the creation of Medicare and Medicaid—all once dismissed as "socialistic." By the 1960s, a host of thinkers were convinced that things would continue on this progressive path until the present day, when all of us were supposed to be working two or three days a week and having our health care provided by the state. Retiring by the age of forty-five was supposed to be the rule by now, not the exception.

These developments in the economy were to spill over into the other aspects of life, where other hierarchies were supposed to come tumbling down. By now students were to be learning at their own pace, gently "guided" by their teachers (if they even had teachers). Children, meanwhile, were to be reared in communes, by the state or even within the confines of polygamous marriages. The '60s spirit of experimentation was expected to still be alive and well right now.

And with all that time on their hands, Americans were going to discover novel, even bizarre ways to amuse themselves: weekend jaunts to Antarctica, say, or watching horse races with robotic jockeys—which could even be more amusing while smoking marijuana, a completely legal recreational drug by the early 21st century.

Not all was sweetness and light. Many of Bellamy's contemporaries believed in eugenics, the idea that "inferior" races would die out, leaving nothing but people of "superior," northern European stock. Later on, observers focused on the dark side of the 1960s—the riots and violence—and concluded that the state would have no choice but to tame its miscreants using cutting-edge mind control techniques.

Far more people, however, thought that the early 21st century would be one in which there would be little distinction between the government and the business sector. A generation ago many thinkers were convinced that Bellamy was absolutely correct.

Prediction: Workers of the World Do Unite

Capitalism as it was practiced more than a hundred years ago made a very few people fabulously rich, some people sort of well-off, and a lot of people overworked and underpaid. With America and other industrialized nations racked by strikes and Karl Marx catching on among the intellectual elite, a lot of thinkers figured that we'd all be comrades by now. For several decades communism was considered the wave of the future—and not just by the fringe.

In the late 19th and early 20th centuries, before there was a "safety net" in the industrialized countries, people who couldn't get work could easily find themselves homeless, hungry, and with no access to medical care. Because of this, worker unrest was becoming violent; in America throughout the late 1800s and early 1900s strikers frequently armed themselves as the hated Pinkerton guards marched in to break up work stoppages and protect so-called scabs. Employees in one industry began staging strikes in sympathy with the workers in another industry, and in a few cases, whites marched alongside blacks and men supported their sister laborers. America, many believed, was cleaving into a two-tier society—a tiny coterie of oligarchs at the top and a huge, angry mass of laborers at the bottom. Many socialist theorists were sure that sooner or later, workers would have had enough and would crash the mansion gates—with grisly consequences for the people inside.

Edward Bellamy was different. A student of socialism and the son of socially conscious parents, he believed that capitalism would gently fold under its own weight long before anyone stormed the barricades, a vision he outlined in *Looking Backward*. In it his hero, Julian West, awakens after a slumber of more than a century in a Boston nothing like the savage, dog-eat-dog world from whence he came. He is revived by a Dr. Leete, and the rest of the book takes the form of a dialogue between the two (as well as other characters) about the modern, 21st-century world. During their conversations Leete proceeds to shatter every one of West's assumptions about economics and human nature.

Leete explains to the initially dubious West that early in the 20th century, businesses gradually became bigger and bigger until they formed a megatrust, a single corporation that encompassed every commercial enterprise. When this happened, Leete explains, "the obvious fact was perceived that no business is so essentially the public business as the industry and commerce on which the people's livelihood depends," and entrusting such a vital societal function to self-interested capitalists made as little sense as letting kings and nobles dictate political life. With that realization, the state took over responsibility for producing all goods and services, not in the name of profit but for the betterment of all.

In this world, citizens were issued something like a credit card, which was worth a person's "fair share" of the nation's economic output. When you needed a pair of shoes, you would go to the government warehouse and the value of the item would be deducted from your card. The government's surplus wealth—and Bellamy believed this system would produce massive surpluses—was used for the "adornment" of the country's public places.

In this world, greed was not good—in fact it was nonexistent. With all material needs guaranteed to be met, people were no longer motivated by the acquisition of wealth. Instead, high achievers sought the recognition and respect of their peers. Especially productive people were awarded military-style badges and medals, on the theory that every-one, not just soldiers, really craved glory. The desire for huge sums of money under the capitalist system masked a more profound craving for admiration, Bellamy believed. Under this socialized system, that deeper thirst for glory would be met, and everyone would benefit.

Naïve? Certainly the engineers of the Russian Revolution didn't think so, nor did some serious political thinkers in the West throughout the 20th century. In 1904, for example, Eugene V. Debs, a Socialist Party candidate for president, proclaimed that capitalism was "dying. . . . The time is near when the cadaver will have to be removed and the atmosphere purified." John Maynard Keynes, writing in 1931, briefly turned his mind away from the depression then afflicting the globe and toward the world he believed would be emerging in the early 21st century, one where technological progress meant that no one would any longer be forced to work for a living. "Love of money," the famous economic theorist wrote, "will be recognized for what it is, a disgusting morbidity." Usury and making money by lending money would no longer be a chief driver of the economy. People would choose to work jobs that provided the most spiritual satisfaction, not the fattest paychecks.

Keynes was and still is widely respected (even Richard Nixon, not exactly a bleeding-heart liberal, was a self-proclaimed adherent) so it's no surprise that Keynesian thought dominated the postwar period. A physicist, Boris Pregel, predicted in the 1940s that abundant nuclear energy would make goods so cheap to produce that there would no longer be a money economy and only a few people would still attach value to the possession of material things. In the 1960s, Nicholas Rescher believed that the Protestant work ethic would give way to a service ethos, one where most people put a premium on helping others instead of "getting ahead."

Around that time, theorist Hans J. Morgenthau predicted that the government "will provide all with the necessities of life," shifting the "gravity of power from the private to the public sector." Others, meanwhile, believed that the U.S. government, faced with millions of "surplus" workers left jobless thanks to automation, would resort to New Deal–style public works projects in order to achieve full employment. Some even believed that America would also have too many skilled workers, forcing the United States to export technicians, doctors, and other professionals overseas—quite a reversal from what actually happened, a world in which America imports foreign workers to do scientific research, program its computers, and tend to its sick.

Still others believed that the government would provide all with a guaranteed minimum income—no matter what they did or even if they didn't work a job at all. And a host of prognosticators foresaw the expansion of the planned economy, one where taxes collected by the government, not private investment, would provide the capital for industry.

Other 1960s-era forecasters, meanwhile, predicted that the United States was at a minimum bound to follow in the footsteps of the Europeans and Canadians and offer universal health and other kinds of insurance. Dr. Oscar Creech, writing in 1967, foresaw a time when the nation's doctors would become federal employees.

Massive state intervention in the economy, however, would actually fall out of favor in the next few decades. In America, President Ronald Reagan slashed taxes, deregulated industry, and promised to get government off people's backs, not expand its role in citizens' lives. Reagan's British counterpart, Prime Minister Margaret Thatcher, also worked to dismantle government controls on the free market. China embraced capitalism, and other nations privatized or sold off their state industries.

Instead of a world where income was redistributed, today, especially in the United States, there has been something of a return to Gilded Age levels of income inequality. Nobel Prize–winning economist Paul Krugman has noted that the "material abundance" predicted by Bellamy, Keynes, and others went mostly to a tiny wedge of the American populace. From 1975 to 2004, for instance, the average college graduate saw his or her income rise about 34 percent, but the incomes of those in the 99th percentile rose 87 percent during roughly the same period, while the very, very wealthiest—those at the 99.99th percentile—saw their incomes increase by almost 500 percent. While today's America would largely be unrecognizable to Bellamy, he would certainly be accustomed to, and saddened by, one in which most of the nation's money was going to just a fraction of its people.

"Socialism," "nationalization," and "regulation" are once again in the air, of course, following the epic collapse of the economy in the fall of 2008. The federal government, just like in the New Deal days, is creating jobs to get the millions of unemployed working again. Under President Barack Obama, the United States is reweaving the safety net that had frayed somewhat in the previous few decades.

So will Bellamy be proved right, after all? Will we all one day become federal employees, living in harmony in a world where our money has been replaced by a ration card? Don't bet on it. The measures America and other countries under-took in 2009 to kick-start the economy were intended to get us through a rough patch, to fix capitalism, not kill it. None of the officials who coped with the Great Recession, unlike many of their counterparts during the Great Depression eighty years earlier, were trying to create a socialist utopia. Today, greed still has its place.

Prediction: The Five-Day Weekend and Forty Years of Retirement

Sometime in the last 150 years or so, the sabbath, the day of rest, became the weekend, two days of rest. It was during this time that the vacation—the truly radical idea that the boss should pay you even if you're not working—also emerged. So too did the concept of the government pension, where you continued to get a check once you got too old to clock in every morning. The official workday used to be sunup to sundown, then became twelve hours, then ten, until we finally arrived at the eight-hour norm. Since these concepts evolved gradually, a lot of thinkers believed that they would keep on evolving, that, slowly but surely, more and more of the company's time would become our time.

By now, it was thought, even accountants and construction workers would be enjoying five-day weekends, not just party-hearty frat boys. Or we'd be retiring while our hair was still, mostly, free of gray. Or we'd be spending as many days, if not more, on vacation as we did on the job.

Things, obviously, have not quite worked out that way. In fact, many younger workers today might actually wind up spending more of their lives working than their parents and grandparents did.

The belief that progress would allow us to work less and earn more had been percolating throughout the 20th century. Eighty years ago, for instance, Keynes believed that as it became easier and easier for man to meet his material wants through technology, what work that remains to be done in society would have to be "spread thinly"—so thinly that the standard workweek would drop to about fifteen hours by the 21st century. In fairness to Keynes, something like this had already happened during his lifetime. In the United States, during the Depression, the forty-hour workweek was created specifically to give more people a job (two workers each putting in forty hours would replace one worker putting in eighty).

By the 1960s, when computers and automation were starting to make a real impact in the workplace, commentators noted that machines would be doing the work of thousands of clerks and assembly line workers. A RAND Corporation study predicted that by the 21st century, just 2 percent of the population would be needed to produce everything society needed, and only "elites" would be able to find gainful employment.

This "banishment of paperwork," IBM scientist Arthur L. Samuel predicted in the mid-1960s, would have led to the institution of a four-day workweek by 1984. And Samuel was hardly alone. Construction magnate Henry Kaiser predicted in 1963 that in the 21st century, the thirty-hour workweek would be the norm, and most American families would be taking monthlong vacations. In 1967, the *Wall Street Journal* envisioned this century as a time when husbands, thanks to greatly reduced working hours, would be spending more time with their wives (who still had not entered the workforce in appreciable numbers). Orville Freeman, the U.S. secretary of agriculture, believed we would be working a third fewer hours than we were in the late 1960s—yet still earning more money. In the mid-1970s, the U.S. Forest Service predicted that the "weekend" would have to be interspersed throughout the week, so as to take pressure off recreational facilities. Other prognosticators, meanwhile, saw people laboring just two or three days a week, or perhaps only six months a year—and again, still having more than enough money to meet their wants.

And when would one's working life officially end? Long before the onset of old age, many believed. Bellamy, for instance, thought that by now one would begin working at the age of twenty-one and be done by forty-seven. In 1968, city planner William Wheaton believed that people would be earning a lifetime's worth of income in just ten years, and we would spend half of our adult lives as retirees. The historian Arthur Schlesinger, Jr., thought that the forced idleness of workers who were still in the prime of their lives represented a real danger to society.

Forced idleness, of course, turned out not to be quite the problem Schlesinger envisioned. It is true that in many parts of the industrialized world, workers do enjoy several weeks of vacation, as well as other forms of leave. In Europe, for instance, monthlong vacations are not unheard-of, and even in America workers can take additional time off when a baby is born or for a family medical emergency—a perk unknown in the 1960s. And one major industrial nation, France, has instituted a thirty-five-hour workweek. But even there, business has tried hard to roll back that advance, complaining that the move has made French companies less competitive with companies overseas (no other country has sought to alter the forty-hour workweek since France's move a few years ago).

And far from working less, Americans and many others are working just as hard, if not harder, than their parents did. Instead of growing, real income for U.S. households has actually been stagnant, on average, for the last thirty-five years—and this during a time when millions more women have entered the workforce. Instead of a sole breadwinner able to provide for the family on a twenty- or twenty-five-hour workweek, the norm today is a two-income couple putting in long hours just to keep up with the mortgage and save enough for the kids' college education.

And in no part of the industrialized world are workers being put out to pasture as quickly as many experts once thought. The United States has actually raised the retirement age from its mid-1960s level and few young American workers expect the government to provide for them in their old age. European leaders, meanwhile, have repeatedly pointed out that on current trends, there will be too few workers supporting too many retirees—unless the mandatory retirement age is revised upward.

Technology, meanwhile, has had some unintended consequences. As the experts predicted, machines have lightened the load; computers now perform a lot of the repetitive tasks that used to require an armada of office drones. But modern telecommunications—the e-mails, the instant messaging, and all the rest—means that we don't spend as much time truly "away from the office" as people did in the days before telephone answering machines. Instead "the office" now comes home with us, in ways that forecasters didn't foresee back when the four- or five-day weekend seemed just around the corner.

Prediction: School's Out—Way Out

Fifty years ago two of the sweetest words in a kid's vocabulary were *summer vacation*. Today two of the sweetest words in a kid's vocabulary are—*summer vacation*.

Progressive educators a generation ago believed that lots of children were ill served by their rigid, rules-based classrooms, that making third-graders endlessly recite their five-times tables was not the best way to prepare them for a dynamic, technologically advanced society. Education theorists back then believed the answer was to stop thinking of students as vessels that had to be filled with knowledge but as partners in a learning adventure. It was time, in a sense, to let the inmates have a little more say in how the asylum was run, making school not only more relevant for the 21st century world but also—yes—fun. Today, it was hoped, kids wouldn't be quite so glum as those last few days of summer vacation drained away.

People who thought about the future of education believed it was imperative to make education more palatable for another reason—we were all going to have to get a lot more of it if we wanted to make it in the modern world. True enough—a college degree today is a lot like what a high school diploma was back in the 1950s or '60s, a passport to the middle class. But many forecasters back then extended the comparison a bit too far. Like a few years of high school are today, educators figured that by now, a few years of college or the equivalent was going to be not only free, but mandatory.

In his 1967 book, *Extended Youth*, Robert Prehoda quoted authorities who believed that by now the norm would be for people to stay in school full-time until the age of thirty, and that we would spend as much of our lives in the classroom as we did working. The *Wall Street Journal* in the late 1960s quoted experts who thought that compulsory education in America would start at the age of four and continue through two years of trade school or junior college. In the late 1950s, the *Harvard Business Review* predicted that by 1970, universal college education would be the norm, the third in a series of historical "waves" that first saw compulsory grade school education followed by mandatory high school.

In a fast-changing, technologically advanced world, required schooling was not going to stop in one's youth, either. To keep their skills sharp, some predicted a generation ago that the middle-aged would also be forced back to the classroom. In 1967, *Futurist* magazine predicted that those in their forties and older would not just be encouraged to go back to school periodically—they would be required to. "The middle-aged drop-out will . . . be increasingly viewed as a menace to others as well as a problem to himself," the magazine predicted. "Within the next twenty years, he may be running from the truant officer."

Back then, the college of the future—and schools in general—were expected to be much less rigid than the warehouses where students were divided up by grade and subject. Instead the walls were going to come tumbling down in the late-20th-century schoolhouse.

Grade school class levels defined by age, for instance, were expected to "disappear." Instead of students getting a passing overall grade based on the average of several subjects, today's twelve-year-old was expected to be studying math, for example, at a tenth-grade level and science at a seventh-grade level, depending on his or her individual aptitudes. A University of California professor and expert on education, Robert Bickner, suggested that today teachers of young children would be less obsessed with imparting specific knowledge and would no longer be grading everyone "on a common scale." "[M]ight we even lose our compulsion to thrust the same pre-digested diet of knowledge down the throat of every student and press every student into a common mold?" he wrote in the late 1960s.

At the university level, meanwhile, many of the trappings that have defined higher education in the West for centuries were also to be scrapped entirely. Not just grades but univer-sity departments were to "whither away," claimed Hendrik Gideonese, an official with the U.S. Office of Education, in a 1968 interview. A few years later, in the early 1970s, futur-ist Alvin Toffler thought it was possible that "long before the year 2000, the entire antiquated structure of degrees, majors and credits will be a shambles." Writing in *Look* magazine in 1967, Marshall McLuhan, the renowned author and thinker, wondered whether students would still even be "graduating" from college at all.

In that same article, McLuhan summed up the general attitude toward the future of education as many progres-sive theorists at the time saw it, one where traditional pub-lic schools—designed to prepare millions for assembly-line work—gives way to a system where the primary function of school is to create adaptable freethinkers capable of life-time learning in a new, modern economy. "Teacher-led" in-struction would give way to a system where students "move freely from one activity to another" within a campus. "Fixed schedules and restrictions on students' movements" would be deemed "unnecessary." All of this, it was believed, was going to produce students far more intelligent than any who had come before. One commentator predicted that the average student of 2000 would be considered on a par with a genius of 1970.

In fact, at least in America, academic performance is, by many estimates, worse today than it was a generation ago—and it's certainly not true to say that the typical high school junior in 2009 would be considered as bright as the brightest of forty years ago.

Nor has a university education become compulsory, although many more people attend college today than in the late 1960s. Unfortunately, however, the progress America was making toward providing all with higher education has been uneven. While more high school graduates do attend college, the portion of students finishing high school appears to have peaked in the 1970s at around 75 percent, a figure that has remained virtually unchanged ever since. And while some innovative teaching methods have been introduced, the classrooms of today still consist of students at well-defined grade levels learning a preset curricula. (Indeed, a common complaint among modern teachers is that they are forced to "teach to the tests," making sure their students master a very specifically designed body of knowledge.) College students still major in a particular course of study and are still required to earn 120-odd credits to earn an undergraduate degree.

For all the changes that were predicted forty or fifty years ago, though, no one appears to have foreseen one of the biggest: the rise in homeschooling. Hundreds of thousands of American students today are getting precisely the type of education the progressives were recommending back then: highly personalized, often loosely structured, designed to allow the young learner to proceed at his or her own pace. And, irony of ironies, many of the homeschooled come from religious, conservative families—the other end of the cultural spectrum from the '60s intellectuals who largely celebrated the freewheeling, hippie ethos. Few back then figured that the classroom of the future would be the kitchen table, or that many of the teachers would begin the day's lessons with a good old-fashioned Bible reading.

Prediction: Knowledge Pills—Swallow and Learn

The attempt to supply smarts in a bottle started in an unlikely way: with a type of worm called a planarian, which practices cannibalism. In the early 1960s, a University of Michigan researcher trained specimens to avoid an electric shock. The researcher then introduced some untrained worms into the same tank; they promptly made a meal out of their learned cousins. The eaters were then able to learn the same trick much more quickly than the worms were normally able to. The scientists came to the conclusion that the worms had quite literally digested their lesson.

If knowledge can indeed be "eaten," it follows that memories are stored in some kind a molecular form, possibly either as proteins or as RNA, the compound similar to DNA that carries out the genetic functions of the cell. With the general concept in place, many scientists sought to isolate the substance "containing" memories—including, some hoped, memories of learning, say, calculus or high school French.

And where memories can be created, some scientists suggested, others can be wiped out. Conducting another version of the Michigan experiment, untrained planarians were fed "smart" planarians along with a chemical that destroyed RNA, the substance some believed played a role in storing knowledge. Sure enough, the brain food didn't work on the planarians this time.

While several scientists were able to replicate aspects of this work, memory/knowledge transference has never been routinely repeated in the lab, the standard scientists apply when determining if an experimental result is valid. But that didn't stop the speculation.

Robert C. W. Ettinger, the proponent of cryonic freezing discussed earlier, thought that it might one day be possible to re-create someone's personality simply by "reading" the chemically stored memories located in a single cell. With just a fragment of a deceased person, you might be able to reconstruct that person's experiences and even personality quirks, a potential boon to historians who might want to know what Napoleon was really like.

Others foresaw profound consequences for society should memory and knowledge be linked to specific molecules. Science writer Gordon Rattry Taylor, while doubting that learning could ever be "eaten" (he didn't think that the knowledge would fare too well in the human digestive tract), did hold out the possibility back in the 1960s that learning could potentially be surgically implanted. The world he envisioned was one where the operating room replaces the classroom. Athletic competition would become redundant (with a dash of Kobe Bryant anyone in reasonably good physical condition could star in the NBA Finals).

Another researcher, brain surgeon Wilder Penfield, had come to different conclusions about how memories are stored in the mind, but his speculations were equally fanciful. In the '40s and '50s, Penfield noticed that as he operated on patients who were not completely anesthetized for certain operations, they would spontaneously recall events from their pasts when his instruments made contact with certain parts of their brains. Moreover, Penfield reported, those memories were vivid—they included recollections not just of sights but also of sounds and smells. Rather than a chemical tape acting as a repository for the past, Penfield believed certain areas of the brain were the attics where we kept all of our recollections. Using electrical stimulation of those specific brain areas, Penfield wrote in 1967, in the 21st century a "perfected technology of memory playback will . . . completely falsify that one famous eternal verity, the old saw that 'You only live once!'" A ninety-year-old in a nursing home would be able to take a "vacation" back to his first kiss; a second honeymoon could be a mere replay of the first.

A lot of scientists had reservations about the possibilities of identifying discrete chemicals or brain areas where knowledge was bottled up. They were especially dubious about claims that skills or information could be packaged like medicine or a treatment. But even some of those very same scientists were optimistic about the possibility of some kind of brain "booster," a way to make people learn faster and more efficiently.

While it has never been shown to be the substance in which specific knowledge is stored, RNA has repeatedly been demonstrated to enhance rats' ability to learn a maze, sometimes up to four times faster than untreated specimens. In other animal experiments, scientists dissected the brains of the smartest individuals and discovered elevated levels of certain compounds, leading some to conclude that a concentrated shot of the right cocktail could potentially add a dozen IQ points or so (the question of how long the effect lasted always remained a subject of some debate, however).

A half century ago, French scientist Jean Rostand believed that by now these substances would have been nailed down and identified. "Just as at present a diabetic gives himself insulin every morning to regulate the supply of sugar in his blood, so perhaps in the future will everyone give himself an injection of something or other to make himself even more intelligent."

Starting in the late 1970s and continuing right up until the present day, a slew of drugs have been discovered that help some senile patients restore part of their intellectual functioning, while other drugs have helped students with hyperactivity disorders concentrate. Scientists have also learned that stimulants such as caffeine and nicotine, whatever else their other drawbacks are, do seem to make users more mentally acute.

But all this falls well short of the wildest hopes of researchers from the '60s, '70s, and '80s. The magic compound that would "cure" mental retardation has yet to be found, even though some believed such a breakthrough would have happened by now. There's no shot, potion, or treatment that today's typical C student can take that will suddenly help him rack up a perfect score on the SATs. Nevertheless, the search continues: in 2007, Foresight, a British government think tank, issued a report claiming that "cognitive enhancers" could be as "common as coffee" sometime in the next few decades. The report also envisioned a time when students would have to be screened for performance-enhancing drugs before taking exams, like Olympic runners before a meet.

Prediction: Family Ties—Unbound

A lot of people place the "women's lib" movement firmly in the 1970s, but in reality the ladies began asserting themselves long before. By the 1920s, in America and elsewhere, women had the vote, were going to work in greater numbers, were partying alongside the men in what had formerly been male-only taverns, and—gasp!—were even wearing pants. Other women were opting for an androgynous style, cutting their hair short and wearing ties and sport coats. They flew planes and drove cars, too.

Observing all this, a certain type of man back then figured that society was on the brink of collapse. As women became more assertive, a lot of men feared, they would forget their "natural" place in the order of things, wrecking the family—and as the family went, so too would civilization. Male thinkers at the time, speculating about what the present would be like, were horrified.

Perhaps none were as alarmed as an author and popular conservative polemicist named Anthony Ludovici, who published a book in 1925 titled *Lysistrata, Or Woman's Future*, in which he predicted that in the decades to come these burgeoning liberated women would become Amazons who would dominate society. And since women weren't suited to lead the world, Ludovici believed, they would of course make a mess of things.

According to Ludovici, industrialization meant that women could now be economically competitive with men in ways they could not be in an agrarian society, where physical strength counts for more. He believed that eventually, due to the male's "besotment," women would take over industry entirely. Once males became economically superfluous, women would choose to perpetuate the species not through sexual contact but through artificial insemination—and they would make sure they had mostly girls (men, under this female-dominated society, would not be allowed to exceed five per thousand of population). Society would only be salvaged, Ludovici predicted, when it once again returned to its roots in the soil and women were once more preoccupied solely with breeding.

Ludovici, who died in 1971, lived long enough to see society withstand the likes of Gloria Steinem and the miniskirt. He also lived long enough to see a number of theorists who, like him, thought that science was going to force everyone to rethink woman's roles, including the roles of wife and mother. The difference, though, was that these theorists believed society would be better as a result—if also a lot stranger.

One of the earliest and most influential of these thinkers was B. F. Skinner, the famous Harvard professor of psychology whose views affected everything from the construction of baby incubators to the layout of supermarkets. Skinner was a prominent advocate of operant conditioning, in which good behaviors were reinforced by bestowing or withholding rewards. He also believed that the family, as it had been known throughout human history, was in desperate need of an overhaul. All the hang-ups and neuroses most of us acquire while growing up in traditional households could largely be cured if we took the sentimentality out of child rearing and replaced it with science.

In his 1948 novel *Walden Two*, Skinner describes an ideal community of one thousand people, in which, among other things, the "primitive" concept of a family defined by blood lines has withered away. Instead parents are discouraged from showing their own children any special favor, and the kids are urged to refer to them by their first names. Children are tended to by specially trained professionals (men as well as women) who are compared to "skilled laboratory technicians." "Even when the mother knows the right thing to do, she often can't do it in a household which is busy with other affairs. Home is not the place to raise children," one character explains.

In this community, the village really would be raising the child. One's biological parents were just two among the many adults who would shower the child with affection. Instead of being saddled with an incompatible "parent" through an accident of biology, older children would seek out adults with whom they had more in common. Children who never wanted for love and who bonded with a variety of different adults would grow up to become more psychologically healthy and well-adjusted than any generation before, Skinner argued.

Skinner's ideas were to have a profound influence on the commune movement a generation later, places where idealistic young people tried to put at least some of *Walden Two*'s ideas into practice. In the late 1960s, communities like the Walden House in Washington, D.C., and Twin Oaks in rural Virginia were formed along explicitly Skinnerian lines. According to its residents, the "family" was no longer a social and psychological unit. The community dined together but kids weren't necessarily seated near their parents. One resident, Joe Anuszkiewicz, believed that the rest of American society was bound to follow Twin Oaks' lead eventually.

In actuality, few communes ever adhered strictly to Skinner's ideals. Although his ideas about child rearing were radical, Skinner was fairly conventional when it came to sexual relationships between men and women. He believed in sexual equality and strongly advocated platonic friendship, but he also believed that rational behavioral science would actually lead to stronger monogamous relationships than were seen in the wider world. Communes, by contrast, became notorious for "free love"—something that Skinner did not especially approve of—where the very idea of the romantic couple was discarded. Anyone could sleep with anyone else, and nobody would ever get jealous (that, at least, was the theory).

The love-beads-and-face-paint set weren't the only ones seeking to tinker with the idea of the family headed by a longtime married couple—partly because by the late 1960s, it seemed as if society would have no choice. Both the divorce rate and out-of-wedlock births were rising and set to go higher (this despite a mid-1950s forecast by the American Sociological Society claiming that the U.S. divorce rate would be below 20 percent by the dawn of the 21st century).

In *Future Shock*, Alvin Toffler wrote that in the future, marriage—like so much else in society—would widely come to be seen as a transitory state; men and women would enter into committed relationships knowing that they would likely be "short-lived," an institution Toffler referred to as "temporary marriage." Serial marriages—people getting hitched three, four, or five times in their lives—would be commonplace. Toffler even noted that in some quarters, the concept of a "trial marriage"—an official union but short-term by design—was gaining traction among some thinkers. Toffler also foresaw young adults leaving their parents' homes even sooner than they had a generation earlier, with millions of people independent and into their second marriages by their mid-twenties (quite a contrast with the global phenomenon seen today, where a number of never-married children stay in the parental home well into their late twenties or even their thirties).

The underlying problem that serial marriage was meant to address, Toffler and others pointed out, was the unlikelihood of anyone meeting their "soul mate" in their youth, someone with whom you could share your life for four or five decades. Another widely discussed solution to this problem was the ancient Middle Eastern and Mormon practice of polygamy, or plural marriage.

As far back as the 1920s British writer Norman Haire believed that legally accepted concubines would make a comeback, given the male's supposedly innate tendency to cat around: "At present we pretend to be a monogamous people in spite of widespread fornication and adultery. . . . But sooner or later we will have to drop the pretence and admit that men are polygamous. . . . Surely it would be better to give all the women half a husband . . . than to give half the women a whole husband and the others no husband at all."

By the 1960s, this general idea was more popular than some might expect. Robert Rimmer, a Boston businessman, wrote *The Harrad Experiment*, which earned something of a cult following among its two million (mostly young) readers. Rimmer, influenced by the introduction of the birth-control pill a few years earlier, wrote about a fictional college where students were actively encouraged to sate their prodigious sexual appetites (physical education classes, for instance, were held in the nude) with as many partners as they wished. In later novels Rimmer expresses the view that the monogamous union was like "solitary confinement," where middle-aged couples are imprisoned in a deadening existence. Rimmer even believed that more, and more varied, opportunities for male sexual release would help bring about world peace and brotherhood.

Unlike traditional polygamy, however, Rimmer was not calling for the creation of harems but rather truly plural marriages consisting of three or four couples, ones where the women had multiple partners, too. In Harrad, for instance, "Valerie," a member of a group union that calls itself inSix, describes how much she enjoys having three husbands, each of whom takes turns of one week sleeping with her: each week, she says, "I feel like a new bride all over again."

Shifting mores were to be just one of the great shapers of the future family. The other was technology. As we saw earlier in this book, advances in genetic engineering, the creation of an artificial womb, and the ability to implant one woman's egg into another woman's womb were expected to wreak havoc with the very concept of parenthood. In a lengthy article published in *Mademoiselle*, a woman's magazine, in 1966, psychiatrist Hyman G. Weitzen warned that the coming "biological revolution" might leave women bereft, deprived of what Weitzen described as their primary creative outlet—the production of children. Denied this supreme joy of creation, alienated from a child whose genetic traits came not from her and her husband but were created in a lab, Weitzen believed that women "may be driven to outdo man" in the professional sphere, "thus reducing him to a psychological nonentity." The male—"who no longer builds his own house, hunts for his meat or makes repairs that demand physical labor"—would likewise be denied the elemental pleasure of at least seeing his genetic inheritance passed on to the next generation. Weitzen wondered whether technological advance might "destroy man's ego totally," his personality rendered "obsolete."

Things, of course, have not gotten quite that bad. Predictions about the death of motherhood seemed to have ignored the fact that even a generation ago, couples were happily adopting children with whom they shared no genetic links.

Contrary to what Weitzen and others fretted about, dads appear to have retained their relevance, too—even as women have stormed the workplace and are rapidly approaching parity in such professions as law, business, and medicine.

As for radical responses to the spiraling divorce rate—'temporary" marriage, polygamy, etc.—many of today's young adults have largely opted to marry later in life, a trend that appears to have strengthened the institution of marriage in many countries. Having been reared in dysfunctional unions, or having known peers who were, the children of the 1960s, '70s, and early '80s seem to be trying to avoid the mistakes their parents made. The thrice-married are still the exception, not the rule.

Prediction: That Vacation Is All in Your Head

With a guaranteed income, the government paying for everything from schooling to medical care and machines doing most of the work anyway, people were expected to have a lot of free time on their hands in the 21st century. Many believed that this unprecedented world where most people would be spending their time playing would require a massive shift in outlook and personality.

Back in the 1950s, the scientist A. M. Low envisioned a time when natural workaholics might be compelled to spend some time just goofing off, "a weird but not absurd future in which drab men are marshaled into dreary sports grounds to have their compulsory entertainment." A decade later, political scientist James Allen Dator believed that in a leisure culture, people would be forced to reassess their ethical codes entirely, abandoning rigid concepts such as "right and wrong" or "sacred and secular."

The futurist Herman Kahn, meanwhile, believed that the emergence of a leisure society would be disastrous for Americans, a "vocation-oriented" people. One American in ten, Kahn thought, would adopt a "beatnik" sensibility, spending their lives unwashed and high on drugs. In traditional Europe, by contrast, the "gentleman" would make a comeback.

And what would we actually be doing with all this free time? According to one forecast from the mid-1920s, not going to the movies. "[T]he speaking films bear within themselves certain weaknesses which may prevent them from gaining great popularity," E. E. Fournier d'Albe wrote in his book *Quo Vadimus?*

Not just movies were going to be out of fashion. So were football, baseball, basketball, soccer, hockey, and anything else that required "the balls and clubs of our ancestors," Low predicted sixty years ago. Sports that involved physical feats, such as running or "knocking a ball in a hole with fewer strokes," would come to be seen as "quaint, childish or brutal." Instead, high-tech sports—high-tech by the standards of the 1950s, anyway—would be filling the airtime on ESPN. People would be tuning in to watch races between radio-controlled planes. Robot jockeys would replace the human variety. (Although he was wrong about people spending the day watching robo-jockies or at the model-plane races, in fairness to Low, video games are a rough approximation of what he was imagining.)

Boxing, Low believed, would stay, but its brutal aspects would be tempered by technology. Pugilists would wear special suits laden with electronic sensors that recorded every blow and kept score, making the sport "much more dependent upon skill and remove the need for physical damage." By now, spectacular knockouts were to have been knocked out.

This was also to be a world that was supposed to have high-speed, inexpensive planetary transport, one where virtually any spot on the globe would be suitable for a weekend getaway—even some destinations that, by today's standards, are still pretty far out. D'Albe, writing more than eighty years ago, believed Western Europeans would pay "an afternoon visit" to Timbuktu or Mount Ararat. People would also be vacationing in the Arctic and Antarctic, taking high-speed jets to get there (little was said about what people would actually do once they got to the poles, however). In 1974, the Forest Service predicted that vacation trips of five hundred miles one way would be routine, happening every weekend.

People were also expected to be taking vacations that were, literally, all in their heads, thanks to various methods of manipulating the brain's pleasure centers. In their mid-1960s book *The Year 2000*, for instance, the futurists Kahn and Wiener predicted that it was likely that dreams would be stimulated, planned, or even "programmed." In this era, people thought that science would be able to conjure up pleasurable experiences simply by tweaking the appropriate brain region or providing a dose of the right drug.

These developments led forecasters such as Toffler to conclude that today, game-show winners may be rewarded by ten minutes of brain "pleasure probing." When you went for a haircut, a device would direct electronic waves directly to the brain, "tickling" a person's fancy. A "psych-corps" was going to be working directly with mental health professionals to produce simulated, pleasurable experiences, "super-Disneylands" where we would be living out all manner of adventurous, sexual, or ego-affirming fantasies. This vision is not unlike that portrayed in the 1990 science fiction film

Total Recall, where vacationers hooked themselves up to a device that piped pleasurable experiences directly into the brain. The hero, played by Arnold Schwarzenegger, plans on visiting Mars and having a fling with a dark-haired beauty without ever having to leave the confines of his futuristic travel agency.

Prediction: For Ladies, No Shirt, No Problem

Like almost everything else—from furniture to dinnerware to spouses to houses—many believed that in the 21st century, clothing would be just one more throwaway, gone after just one wear. From the mid-20th century forward, experts believed that science would give us cheap yet sturdy alternatives to a permanent wardrobe, freeing us all from the hassle of storing and, especially, cleaning our shirts, pants, and dresses. Other experts thought that in a time of scarcity, natural items such as wool would simply be too expensive to use in mass-produced clothing, forcing us to turn to disposables.

In the 1950s, American Cyanamid, a chemical company, developed paper doused with compounds that increased its "wet strength," meaning that it did not come apart in water. First developed for use by the army—which needed waterproof but lightweight maps—the paper, researchers soon realized, could be used for bathing suits, too, as well as dresses. A decade later, in the 1970s, magazines carried ads for paint-yourself paper dresses that sold for as little as two dollars—rivaling the cost of dry-cleaning. Paper clothing was thought to be especially attractive to the parents of young children, who could make a mess out of their outfits without consequences. Even wedding dresses were going to be made out of paper.

Others believed that clothing would be even more durable and convenient than ever before, thanks to advances in the laboratory production of fabrics. Back in the 1920s, commentators believed that the future wardrobe would be so lightweight that half a dozen changes of outfit could be carried in a handbag. In the 1950s, when chemical companies were having great success producing synthetic fabrics, forecasters believed we would all be wearing outfits that were not only stain-resistant but fireproof, too. Shoes would never scuff, and muddy, sticky children's clothes could be cleaned with just the application of a cool stream of water.

One's future wardrobe was also going to enhance the wearer himself in ways that were unimaginable. Swimsuits, for instance, were going to be designed so that the swimmer could move in the water with the speed and grace of a porpoise. A child who wore one of these special suits was going to swim as well as an Olympic athlete after just a few lessons. Clothing was also going to enhance one's shape—but not using time-honored tailoring techniques such as dark colors, cinched waists, and broadly cut shoulders. Instead, small electronic devices embedded within the clothing were going to adjust the outfit's shape on the fly, making potbellies disappear and masking poor posture.

Not surprisingly, styles themselves were going to change, too; a culture where the old-fashioned mores have been abandoned would be reflected in couture. Edward Kittrich in the 1960s developed a double-pronged high-heeled shoe—footwear that was intended to preserve the elegance of the classic heel but also help a woman retain her balance as she went through her busy day on the job. Another designer believed that future ladies' outfits would be adorned with special projectors that broadcast images of butterflies around the wearer—a 21st-century update to conventional jewelry.

This is relatively tame compared to the fashion future sketched out by avant-garde designer Rudi Gernreich in a 1970 *Life* magazine article. In it, he predicted, the wardrobe would have to become a lot more functional to accommodate the coming age. So instead of women fussing with their lashes and eyebrows, these would be shaved and removed entirely. As they entered the workforce in greater numbers, women would have to discard makeup and jewelry, dressing "like tomorrow's man." And as for hairstyles, Gernreich predicted that bowl cuts would be huge for both genders.

Gernreich was right that androgyny was going to be fashionable in the future, but he may have overstated the case a bit. Women would wear traditional men's clothes, he believed—but men were also to be donning skirts on occasion. There would no longer be a sexual distinction in dress. Metrosexuality hasn't gone quite that far yet.

Like a lot of other forecasters we've met, Gernreich also believed that trends that were just emerging in his day would merely continue into the future until they reached some sort of logical conclusion. And since the style trend back in the swinging '70s was more skin (plunging necklines, miniaturizing miniskirts), Gernreich concluded that things would get to the point where nothing was left to the imagination. Gernreich predicted that going topless (as weather permitted) would be socially acceptable for both men and women (although women would likely wear protective pasties). Since "there won't be any squeamishness about nudity," he said, see-through clothing would be in vogue, too. Think about that the next time you're walking down a crowded street.

Prediction: There Will Be a Safe Cigarette

In just the last forty years or so, we've gone from being able to smoke in hospitals and at the office to seeing the demon weed tobacco banned from the inside of bars. The Marlboro man has been knocked off his perch on billboards. In some states, kids get the right to vote, drive, and serve in the military well before they can buy a pack of cigarettes. As a cigarette ad once famously proclaimed, we've come a long way, baby.

A generation ago people were just becoming aware that a cigarette is about as safe, in the long run, as driving at night without your headlights on, but the initial response didn't involve a ban. Instead many thought that, thanks to science, we'd still be puffing away with the same abandon as the old '40s movie stars—but with a crucial difference.

After decades where the American medical establishment officially remained neutral on the question, the U.S. surgeon general issued a bombshell in 1964: cigarettes cause lung cancer. The formidable tobacco industry immediately fired back, conducting its own experiments to counter the mountain of evidence showing how harmful smokes were and promoting smoking's supposed role as a stress reliever. Millions were pumped into savvy advertising campaigns. And the industry also tried to develop a product that somehow didn't deliver nasty, cancer-causing stuff into the lungs. The quest for the safe cigarette had begun.

Beginning in the mid-1960s, researchers started analyzing the hundreds of chemicals in cigarette smoke with an eye toward figuring out which ones did all the damage. The industry also popularized cigarettes with filters (initially a tough sell among fellows who believed they "sissified" their manly habit) and low-tar brands. Forty years ago some manufacturers even toyed with cigarettes made of lettuce, presumably on the theory that the basic ingredient of salad just had to be healthier than tobacco.

All this activity spurred some scientists—many of whom, it was later learned, were in the direct employ of the tobacco companies—to proclaim that the noncarcinogenic butt would soon be here. Just two years after the surgeon general's report, for example, Dr. Kenneth Endicott, the director of the National Cancer Institute, told a congressional subcommittee that it was "technically feasible" to reduce smoking's "carcinogenic hazard." He also told an interviewer that additives, filtration, and genetically modified tobacco together would "inevitably yield a safer cigarette." During the next decade, other studies were conducted supposedly suggesting that the cancerless smoke could be on the market by the 1980s.

The shills with advanced degrees who were peddling this line finally gave up in the mid-1990s, after cigarette makers were successfully sued in U.S. courts. Tobacco executives were forced to admit that they too were fully aware of what literally thousands of studies had shown all along: the safe cigarette was, and always had been, a red herring. Igniting organic material produces hot gases loaded with carcinogens, and no matter how advanced the "space-age" filter placed between the burning tobacco and the lungs, those fumes cause serious damage.

Prediction: Controlling Minds, Eugenics, and Other Dystopic Visions

Charles Darwin's theories about the origin of the species had a profound impact not just on the study of biology but also on the social sciences. In the mid-19th century, when Europeans were colonizing the nonwhite world while white Americans kept black slaves and were slaughtering Indians, the "survival of the fittest" concept came in quite handy. There was, after all, no real moral justification for the British rule of India or for the supposedly freest country in the world keeping millions of human beings in bondage. Unless, of course, white domination of everyone else could be shown to be merely the "natural order." According to many—and not just Adolf Hitler—in the early 21st century the human race was going to be "purified," understood to mean that only "superior" bloodlines would still exist.

Prominent thinkers for a century were quite candid in their belief that northern Europeans were physically disposed to be the masters of the world, the top of a pyramid with Asians in the middle, roughly speaking, indigenous Americans further down, and black Africans at the bottom. Many of these thinkers were convinced that some nonwhite populations would gradually die out, provided "sentimental" governments didn't make any ill-advised attempts to preserve these "inferior" breeds. These attempts to supposedly improve the human race were known as eugenics.

In his *Anticipations*, H. G. Wells spoke for many of his contemporaries when he wrote that in the future, in an idealized New Republic, the "swarms of black, and brown, and dirty white and yellow people" will "have to go": "So far as they fail to develop sane, vigorous and distinctive personalities for the great world of the future, it is their portion to die out and disappear." Jews, Wells also believed, would not die out but would be assimilated; their culture absorbed by the majority Gentile society around them. The writer Norman Haire advocated infanticide of inferior babies and the sterilization of the unfit so society's resources could be focused on preserving higher-quality people.

The divide between the inferior and the superior wasn't just defined by race, however. Many eugenics proponents—including George Bernard Shaw, abortion advocate Margaret Sanger, and Winston Churchill—believed that the poor and uneducated, those with mental problems, low intelligence, or shaky morals should likewise be discouraged from breeding. The great fear among eugenicists was that many of these so-called inferior people also tended to have a lot of babies, babies that had a much better chance of surviving thanks to modern medical care. Meanwhile, the elites tended to have fewer children. Unless something was done, eugenicists feared, mankind's gene pool would be swamped with mediocrity, and the entire human race would degenerate as a result.

Eugenics remained respectable for decades, until World War II, when the world was horrified to see what it looked like in practice in Hitler's Germany. But even in the postwar period, efforts to preserve the "best genes" were still in place. Until the 1960s, several American states had antimiscegenation laws, rules that prevented interracial marriage, on the books. Progressive Sweden, meanwhile, continued to carry out a policy of sterilizing the "unfit"—sometimes defined as people with slightly below-average intelligence or who were just poor—until 1975. And some forecasters believed that in the 21st century, governments, armed with newfound knowledge of human genetics, would actively be working to ensure that only certain people would be allowed to reproduce.

Fearing that intelligence would "continue to degenerate unless something was done," Dennis Gabor, in his 1964 book,

Inventing the Future, wrote that there could one day be a government scheme limiting families to two children each unless the parents could prove they possess exceptional "talents, health, beauty and heredity," in which case they would be allowed to have three or more children. Others envisioned a time when the state would only issue marriage licenses to genetically suitable couples.

Something like eugenics persists today. Genetic testing makes it possible to determine whether parents are the carriers of certain fatal or severely debilitating illnesses, such as Tay-Sachs, and often those parents will elect not to marry. Severely deformed fetuses are sometimes aborted. But for the most part, modern governments don't require people to pass a genetic fitness test before they can wed. In the industrialized world, forced sterilization of the mentally retarded or the mentally ill ceased decades ago. But the state, it was feared, would be intruding in people's lives in other, equally profound ways.

For decades, many thought that one day governments would acquire the means to completely control human behavior, to turn people into easily manipulated automatons. Many believed this would be necessary to cope with a citizenry that was expected to be on the brink of anarchy. In his 1968 book, *The Unprepared Society*, Donald Michael lists a series of possible national crises, from more ghetto rioting to a coast-to-coast strike that cripples the country to an out-and-out race war. In other writings Michael also sketched out the possibility of angry, unemployed youth movements being brutally crushed by police and vigilantes. Such fears revived decades-old talk of government using radical methods to reassert its authority.

In the early years of the Soviet Union, for instance, Lenin envisioned a time when workers, through psychological conditioning, would be trained to perform their tasks with an uncomplaining, machinelike efficiency (a concept mocked by George Orwell in his novel *1984*). Around the same time the philosopher Bertrand Russell believed that by manipulating glandular secretions, state doctors could keep the dominant class vital and aggressive, while the lower class would be kept stupefied and obedient. In Anthony Burgess's novel *A Clockwork Orange*, psychological conditioning is used to tame a violent career criminal, making him so placid that he is no longer able even to defend himself. Twentieth-century technology was seen as the worst enemy of free will.

That fear seemed less far-fetched by the 1960s, when the results of a series of experiments by a brilliant neurological researcher named José Delgado became widely publicized. Delgado developed small electronic devices that could be implanted in the brain and used to, at least in a crude sense, manipulate behavior remotely. Delgado installed his devices in cats, rodents, monkeys—and people.

Perhaps the most dramatic demonstration of this new technology's power came in 1963, when Delgado, while in a bullfighting ring in Spain, pressed a button and a charging bull implanted with one of his "stimoceivers" suddenly slid to a halt. In other experiments, Delgado managed to change an "alpha male" monkey into a submissive weakling who cowered before the other animals he once dominated. And in his experiments on humans—individuals suffering from a range of behavioral disorders—he induced feelings of pleasure or melancholy. He even managed to make a man's hand move despite the man's conscious effort to keep it still. "I guess,

Doctor," the man told Delgado, "your electricity is stronger than my will."

Delgado said that the focus of his research was on the treatment of diseases such as chronic depression and epilepsy, and he repeatedly denied that he was setting out to devise ways to manipulate sophisticated human behavior—he even dismissed such a goal as impossible. Still, Delgado's own statements certainly left room for interpretation. "Functions traditionally related to the psyche, such as friendliness, pleasure or verbal expression, can be induced, modified and inhibited by direct electrical stimulation of the brain," Delgado said in a 1965 interview with the *New York Times*.

The controversy over Delgado's research boiled over a few years later when two of his protégés, Vernon Mark and Frank Ervin, published a book titled *Violence and the Brain*, in which they described their work treating killers and other miscreants using electrical brain implants. Mark and Ervin, like Delgado, also dismissed the possibility of turning people into pliant robots. But they also put their work in the context of the violence of the times, when murder rates were spiraling and riots rocked the cities. While suggesting that traditional social welfare remedies were required to combat this rising tide of violence, Mark and Ervin also said that such methods were insufficient. Pointing out that only a few people actually committed violent acts during the notorious Watts riots in Los Angeles, the two believed that "brain scientists and clinicians can provide significant help." Another researcher tried to use brain implants to "cure" a homosexual man.

Other researchers were appalled at the possibilities. Still, it seemed inevitable to some that such a potent technology would not go to waste (the U.S. Office of Naval Research, for one, worked with Delgado). The futurists Kahn and Wiener, for instance, predicted that in the 21st century, prisons would be employed less, thanks to advanced methods of "controlling" prisoners.

The controversy, however, coupled with advances in the pharmacological treatment of illness, pretty much destroyed interest in controlling behavior via electrical means (although such devices are used today to help paraplegics, Alzheimer's sufferers, and epileptics). Delgado's work, groundbreaking though it was, is rarely cited today. Meanwhile, the U.S. government has officially come out against attempts to short-circuit free will in an attempt to modify behavior.

Mind control was not the only perceived threat to individual liberty in the 21st century. As modern societies strayed further away from their traditional roots, they also appeared to become more fragmented and hence more difficult to control. As a consequence, some foresaw a time when democratic freedoms would have to be scrapped so as to preserve some semblance of order. Bertrand Russell, observing the rise of fascist dictatorships in Europe in the 1920s, believed that concepts such as a free press, free trade, and civilian control of the military were becoming hopelessly quaint. By the 1970s, he believed, the forms of governments would be dictated by "the men who control the armaments and materials."

The threat of a powerful military-industrial complex of the type President Dwight Eisenhower would later describe was a fear among many leftist critics in the 1960s. Michael Harrington worried that advances in electronic data collection and computer controls would lead to a type of corporate feudalism or military fascism. Still others believed that in a surveillance society—one where security cameras were being installed all over the place—the concept of privacy would erode. Even such time-honored arrangements as the confidentiality between doctor and patient or cleric and penitent would disappear. Some believed that even the very concept of individuality was under threat. In 1967, August Heckscher, commissioner of New York City's parks department, presented a paper titled "The Individual—Not the Mass," urging the creation of a commission to preserve individuality. Otherwise, he feared, "It can no longer be expected that if everything is left alone we shall escape being turned into robots."

All of these predictions assumed that America would be a wealthy place in the 21st century. Not everyone, however, was so sure. In 1951, an article in *Changing Times* predicted that the American standard of living was in for a long, steep decline. While convinced that U.S. industry would continue to be wildly productive, the article argued that more of that production would go toward military expenditures in particular and to government in general. The quality of services would decline, "goods will begin to disappear from stores," cost more, or be of lower quality. In 1954, Fairfield Osborn, author of the book *Limits of the Earth*, believed that the American standard of living would not rise "indefinitely" and that a turning point was coming in the relatively near future. A consensus of futurists in the mid-1970s believed that we had then reached Osborn's prosperity tipping point, a time they dubbed the "post-affluent society." This era, those forecasters believed, would be typified by a "new scarcity," in contrast with the period of wealth expansion that America and the world had enjoyed in the three decades following World War II.

As it happened, America and the rest of the world has seen its standard of living continue to rise, although we have to work just as hard, if not harder, to maintain it. The concept of privacy is still very much alive, but has indeed been threatened by even more sophisticated electronic surveillance techniques and even by the probing of our genetic backgrounds (governments have been forced to address issues of genetic privacy). The definition of family is a lot more flexible than it was a generation ago, but most couples still mean it when they say "till death do us part" at the altar. And the advent of the paper wedding dress, even in 2009, still looks like it is a long, long way off. Much as things have changed in the last fifty or sixty years, much has still stayed the same.

Chapter Eight

A GLOBAL PERSPECTIVE: DOMESTIC AND INTERNATIONAL POLITICS

When Soviet leader Nikita Khrushchev reportedly told the West "We will bury you" in the early 1960s, it was, seemingly, no idle boast. The Soviet Union was a fearsome nuclear power that, many believed, would soon outperform the Western, capitalist democracies. Its technology was top-notch; as we saw earlier in this book, the Russians were at one point handily winning the space race, leaving many American educators panicked about the state of the nation's school system. The Soviets also appeared to be winning the global battle for hearts and minds, as a number of former colonial powers aligned themselves with the communists and designed their economies along Leninist-Marxist lines. Unless the unthinkable happened—a nuclear war—many were convinced that at a minimum, the Soviet Union was going to last.

Almost no one foresaw what would actually happen, however: the peaceful disintegration of America's old Cold War nemesis. Almost no one believed that the world would become one dominated by a lone superpower, the United States. Nor did many foresee that China, the world's other great communist nation, would challenge America not in Cold War terms but on its own terms—as a burgeoning free-market state.

No, it seemed far more likely that the nations of the world would come close to realizing a dream that had been developing for centuries: one world government. That seemed a much likelier prospect for the 21st century than the factories of China supplying the United States with everything from pet food to shoes to plastic action figures.

Trying to figure out what the global geopolitical landscape will look like even a few years from now is incredibly difficult—tougher, in fact, than trying to guess about things such as trends in technology or domestic society. Singular events or the actions of a few dynamic individuals are virtually impossible to predict yet can have a profound effect on the world. A successful revolution in Iran back in 1979, for example, changed everything from the world market for oil to the nature of the modern majority-Islamic state. In the early 1960s, North Vietnam's puny military seemed no match for American might, yet it was—and America's defeat had consequences that included massive inflation, the resignation of a president, and a huge loss of prestige for the United States. Or consider a more recent example—the September 11, 2001, attacks on New York and Washington. Fewer than twenty men, with box cutters and rudimentary flight experience, set in motion a chain of events that has included not one but two wars.

As we've seen repeatedly throughout this book, the unknowable or the "unanticipatable" have often rendered even the most sober prognostications comically off the mark. When it comes to politics and international relations, the effects of wild cards are even more pronounced. Many of the people who in past decades tried to describe what the world would be like now painted a picture that was either way too bleak or way too rosy.

Prediction: War Is Over

John Keegan, the British military historian, once challenged the idea that war is a fundamental part of man's makeup, a scourge that has always been with us and always will be, by comparing it to another, seemingly integral institution: slavery. Up until a few centuries ago nearly every society on the planet had slavery in some form or another. Today almost none does. Even though slavery persists in some form in a few places, in no nation is it legal. If the entire human race can reverse thousands of years of tradition by deciding that one person cannot own another, then couldn't we come to the conclusion that killing each other in battle is immoral, too?

Many half thought, half hoped so, and some were bold enough to predict that by now warfare would be a thing of the past. Edward Bellamy, the author of *Looking Backward*, was like many socialists who believed that a world that had done away with cutthroat economic competition would also be one where there was nothing left to fight over. Governments, according to Bellamy, have in the 21st century relinquished their war powers, instead concentrating on fighting the "common enemies" of cold, hunger, and illness. The citizens' martial impulses have been channeled in service to the economic realm.

It wasn't just the leftists who believed that war would become a thing of the past. The wealthy industrialist Andrew Carnegie predicted in 1901 that before the end of the 20th century, "the earth will be purged of its foulest shame, the killing of men in battle under the name of war."

Others believed that the weapons of war had become so lethal that it would be madness to pursue any path but peace. John Jacob Astor IV, the inventor, wealthy heir, and one of the more famous people to go down with *Titanic*, predicted in 1903 that the airplane would usher in an era of harmony, describing its invention as a "happy dawn for earth dwellers, for war will become so destructive it will probably bring about its own end."

The philosopher William James, the brother of the novelist Henry, also thought that modern war's very effectiveness would render it obsolete. In his 1910 book, *The Moral Equivalent of War*, he wrote that "war becomes absurd and impossible from its own monstrosity" when the combatants are "entire nations" and the technology of destruction equals or surpasses the technology of creation.

James was writing thirty-five years before the advent of nuclear weapons, the most potent technology of destruction yet devised. Midcentury forecasters who believed peace might be at hand by now recycled his point, knowingly or not, when they argued that if humanity were faced with the stark choice of peace or death in a day from a nuclear exchange, the human survival instinct would win out. Dennis Gabor, focused strictly on the U.S.-Soviet rivalry, predicted that should humanity "round this very dangerous corner" (that is, nuclear stalemate), we will have entered the era of "post-historic man," a fortunate creature who lives "for his own happiness" and not the "glorification of his rulers." In 1968, a panelist of forecasters that included members of the U.S. military predicted that nuclear weapons would be banned by 2002 and that there would be regular international arms inspections.

Other forecasters stopped short of predicting a new dawn in human history but did believe that today the world would be more peaceable than it actually is. Clark Abt, an engineer who wrote war-game scenarios for the U.S. military in the 1960s, believed it was possible that there would be a resolution to the Arab-Israeli conflict by now, with both sides cooperating on matters such as managing water resources. (However, in that same series of projections, Abt also thought it possible that both sides would be locked in a nuclear arms race.) Writing around the same time, historian Arthur Waskow, while explicitly saying that true world peace had not yet been achieved, did believe that by 1999, wars would be "non-military" in nature. Waskow thought that we would have all disarmed by now, with an international body specifically set up to keep all nations weapons free. He also believed that countries such as the United States might have transformed their incredible military capacity into, literally, forces for peace, with all of those ships and planes retrofitted for humanitarian missions. "The Strategic Air Command might be stripped of its bombs and turned into a Strategic Cargo Carrying Command for the service of the underdeveloped countries," Waskow wrote. "Perhaps the infantry would be turned into a World Conservation Corps."

What Waskow was describing is today referred to as "soft power," and it is indeed a tool in American (and international) foreign policy. But weapons and weapons systems still proliferate. The Cold War is over, but there are still tensions between nuclear powers (such as India and Pakistan or, for that matter, the United States and Russia). Although the threat of global nuclear war has been greatly diminished, millions have still been killed in combat during the last few decades, sometimes with the simplest weapons to hand (the Rwanda genocide in the mid–1990s was perpetrated largely by machete). International peacekeeping bodies have prevented further bloodshed and international pressure has kept some aspiring tyrants in check, but this process has been far from perfect. War, of course, is still very much with us.

Perhaps it wouldn't be, though, if another widely made prediction had come to pass: the emergence of a global government.

Prediction: The Global Government

The poet Alfred, Lord Tennyson wrote of the time when "the battle flags were furl'd in the parliament of man, the Federation of the World." Tennyson had a lot of company in believing that true global brotherhood would not be achieved until we all lived by roughly the same laws, paid taxes to the same government, and, most importantly, no longer had a "Them" to worry about. Under a world government, them would be "us."

In the modern age it seemed as if all the pieces were falling into place for a true global leadership. Advanced communications and transport made every spot on the planet more accessible than ever before. An empirical, scientific mind-set was taking hold among the elites and more and more average people, too, an attitude that was antithetical to superstition and petty regional chauvinism. When the United Nations formed immediately after World War II, it seemed as if we had taken our first step toward a global rule of law.

Well into the 21st century, of course, jingoistic patriotism and mistrust of other ways of life are still very much alive and well. Tennyson's parliament of man doesn't look like it will be convening within the lifetimes of anyone reading this book.

The rumblings for the creation of a global government began in the wake of the World War I, the most lethal conflict in human history up to that time. Pumped up with national pride, the citizenry of many of the combatant nations marched eagerly to the trenches in 1914. Four years and millions of dead later, a deep skepticism had replaced the flag-waving bluster. This was also the time when President Woodrow Wilson first proposed creating the League of Nations, a forum where international disputes could be resolved peacefully. In the Soviet Union, meanwhile, communist leader and theorist Leon Trotsky believed that class would supersede nationality, that the workers of the world were indeed ready to unite, regardless of any lines on a map.

Writing in 1925, the scientist J. B. S. Haldane wrote that "there are many people, and ordinary, normal, people, who favor the idea in one form or another of a world state." H. G. Wells believed that by the 21st century, a mega-state consisting of what was then the British Empire plus the United States (and those parts of Latin America it dominated) would emerge. This mega-nation would then enter into agreements with what he believed would be the other two great powers (the European and the "yellow," that is, Asian) to create unrestricted world trade, a common currency, and universal laws, "by which the final peace of the world may be assured forever."

By the 1960s, others foresaw the rise of a global state as

virtually a necessity for human survival in a world that could annihilate itself with the push of a button. Forecasters back then generally did not believe that we would actually have a global government by now, but that we would be well on our way—much further along, in fact, than we actually are.

Fritz Baade, in his 1962 book, *The Race to the Year 2000*, believed that countries by now would have surrendered "a large part of their national sovereignty" for the sake of world peace. "No visas, nor even passports will be required," he wrote. There will be the "construction of world-wide political institutions that at least approach the vision of a unified world government." He believed that nations with diverse political and cultural traditions would nevertheless be organized in a "federative" body that would lend each nation a "special, new and invigorating vitality."

Stuart Chase, meanwhile, believed that by now the United Nations—which would have been moved to a neutral island—would be presiding over a globe that had largely been free of conflict for a generation. Competition between nations "is largely confined to advances in the arts, research and the Olympic Games." One's country would now be considered a "nostalgic homeland," not something to which one swore allegiance or had to die for. Nicholas Rescher, a prominent American philosopher, believed that in the 21st century the United States and citizens of the other major powers might well have become less preoccupied with international prestige and more concerned about quality of life. In 1968, one group of forecasters predicted that within a decade, international political parties would emerge.

Others believed that a global language would also emerge in order for the citizens of this new world order to comprehend one another. H. G. Wells, for one, thought that English would be the lingua franca of the day (a prediction that was not far off the mark). Still others thought that a supplemental language would be created, with most of us speaking our native tongues as well as this second tongue in order to do business with our foreign friends. This was the dream of L. L. Zamenhof, who created the language Esperanto in the late 19th century. Esperanto, and indeed no other created language, has ever caught on in a big way, however.

In the 1960s, there were a lot of hard-eyed realists who never thought for a second that the individual nation-state would become passé. Instead, they were sure, America would still be locked in an uneasy stalemate with its perennial foe, the Soviet Union. Even those who had an unshakable faith in capitalism thought that the Russian communists would still be around—and thriving. But as for the other great communist nation, China—well, not so much.

Prediction: The Soviet Union Is Alive and Kicking

The sophisticated Western cold warrior, circa 1975, may have believed that the so-called American way of life was superior to the Soviet one, but that didn't mean he thought the Soviet Union would simply vanish from the map one day. Or if it did, it would be in the most horrific way imaginable—by nuclear war, which would almost certainly mean the destruction of the United States as well.

Thirty-five years ago the Soviet Union seemed to be doing fine. The United States may have scored a diplomatic victory by establishing links with the Russians' other great rival, China, but for the most part the future seemed to belong to Moscow, not Washington. The United States had just been

humiliated in Vietnam, the greatest military in the world
beaten by Russian-backed insurgents. America was wobbling
from the Arab oil embargo (the Soviets, by contrast, had
plenty of petroleum of their own). In Africa and Latin America, revolutionaries looked to Lenin and Marx for inspiration,
not Washington and Jefferson. The Soviets were securely en-
sconced in Eastern Europe. Official U.S. foreign policy was
dictated by realpolitik, a philosophy that, broadly speaking,
meant coming to terms with the fact that the Soviets, apparently, were here to stay.

America's own futurists had largely been making the same
case for quite some time by then. In their 1967 book, *The Year
2000*, for instance, Kahn and Wiener believed that in the 21st
century, America and the Soviet Union would still be considered superpowers (trailed by eight other "powers," including
Japan, France, and China). And in some respects, the Soviet
sphere was expected to even surpass the West. The two predicted that the Warsaw Pact countries (Eastern European nations such as Poland, Hungary, and the former Czechoslovakia)
would become more economically powerful than the Western
Europeans sometime between 1985 and 2000. Poland was expected to have a richer economy than Australia by 2000.

Another contemporary forecaster, Ferdinand Lumberg,
looking ahead 150 years, also believed that the Soviet Union
would still be intact, with very little likelihood of the empire
liberalizing. If it were to change at all, Lumberg believed, it
would take a violent revolution, but such an upheaval would
be difficult to pull off in a modern, technologically advanced
state: "Large scale revolution has been made impossible by
modern science," Lumberg wrote.

And in any event, the Soviets of the future would have little to complain about, Lumberg believed—the average citizen was likely going to be fat and happy. Lumberg did point out that in the mid-1960s, the typical Soviet citizen was much worse off than the average American; the former had to work three times longer than the latter to earn enough to buy a loaf of bread, six times as long to buy a pound of meat, and twelve times as long to buy a decent suit. But in the future the Soviet Union was expected to "equal" or "surpass" Western Europe in cultural and industrial achievements thanks to a number of factors. In the fifty years since the fall of the czar, the communists had managed to create an industrial powerhouse from a peasant nation, so there was a precedent for making great strides. The massive Soviet nation, taking up one-sixth of the world's land area, contained huge reserves of coal, oil, and minerals as well as vast growing regions. Its universities were producing thousands of graduates in advanced technical fields.

Fritz Baade was equally sold on the future of Russia. Its system of collectivized agriculture would be sufficient to feed the 350 million Soviet citizens of the 21st century, twice as many people were going to be living in communist regimes as in capitalist ones, and their economic output would be double that of the capitalist countries. "In the race to the year 2000 there are areas where the East will clearly pull ahead of the West," Baade proclaimed.

What actually happened nicely illustrates one of the great pitfalls of long-range forecasting—trying to account for exigencies, or, put more simply, wild cards. In 1980, the Soviet Union got bogged down in its own, Vietnam-like quagmire, in Afghanistan. During that decade, the price of oil dropped, depriving the Soviet Union of a significant chunk of its export income. President Reagan dropped realpolitik and opted for an aggressive program of expanded military spending, forcing the cash-strapped Soviets to try and keep up. Labor unrest in Poland, backed by Polish native Pope John Paul II, inspired millions of oppressed residents across the Soviet sphere. By the time Germans took a sledgehammer to the Berlin Wall in 1989, its foundations had largely been weakened already.

Such unforeseen factors also help explain how at least a few forecasters a generation ago misspoke regarding the future of China, which some believed was still going to be an economic cripple by now. Although hardly a household name in the West, Deng Xiaoping will be remembered as one of the most significant figures of the 20th century, a man who quite literally changed the world. In 1978 he took over leadership of the Chinese Communist Party, a position that allowed him to dictate the future of the nation. But when he ascended to power, Deng's country was a shambles thanks to the policies of his predecessor, Mao Zedong, a doctrinaire communist and tyrant whose Cultural Revolution cost millions of lives. Deng opted for a different path, introducing free-market reforms including the privatization of land and businesses. Although China today is still nominally communist, it has in effect a free-market economy—and one of the most powerful on earth. Many believe that China will once again have the world's largest economy—a position it held through much of its long history—by about 2025. Not for nothing are some experts claiming that we are now living in the "Chinese century."

But when forecasters in the mid-1960s were making their picks about who the major players were going to be right now, no one could know the influence that Deng would eventually wield. Instead they saw a nation of mostly poor peasant farmers who still plowed their fields the way their ancestors had done, an agricultural system that hardly seemed capable of feeding the most populous nation on earth. It was also one where, during Mao's revolution, engineers, doctors, managers, and the other technocrats a modern society needs to function were killed, imprisoned, or stripped of their jobs. China to some seemed a no-hoper.

Lumberg, for one, was particularly sour on China's future. "Red China . . . seems likely to survive, but that China will in 50 years be able to industrialize to the extent that Russia has in a similar period of time seems impossible," he wrote in 1963. Among Lumberg's other wrong guesses about China, he predicted that the nation would be unable to control its birthrate, with the nation crushed by overpopulation (as was noted earlier in this book, China, thanks to its one-child policy, did dramatically slow its population growth). He believed that even by the year 2100, China would still not have caught up to Russia in terms of industrial development, much less to the West (by 2004, China was the world's third-largest exporter and still climbing). And, unless the Americans and Russians had destroyed each other, China would not be a world power, he also predicted. As it turns out, China today is the sole main rival to America's current status as a superpower.

And what would be going on with that superpower? Among other things, some thought it would have more than fifty states and push-button, direct democracy.

Prediction: Puerto Rico Is the 51st State, the Two-Party System Is Finished, and Other Developments

Since before the Revolutionary War, one region of America has been practicing the purest form of democracy of them all: direct democracy. To this day, in towns across New England, residents gather annually to do a job relegated to lawmakers in the rest of the country, voting on things such as budgets and laws. Outside New England, voters partake in representative democracy, meaning people appoint politicians to make the big decisions.

Although having citizens vote directly on those issues might seem ideal, it would also be difficult to carry out in a city of two million, much less a state or on the national level. Tallying up the yeas and nays in the town hall of a New Hampshire hamlet is one thing, but counting up millions of votes on, say, a war resolution is something else entirely.

Unless, of course, the technology existed to keep track of all those votes. With the arrival of powerful computers in the 1950s, some believed that in the 21st century Americans would eventually be participating in *American Idol*–style electronic voting on issues affecting their daily lives.

In the 1960s, futurist Theodore Gordon thought that one day citizens would watch legislative debates on a television channel much like today's C-SPAN. Then, when it was time to vote, the average person would phone in his decision. Physicist A. M. Low envisioned a time when citizens would decide at least smaller-bore issues within the space of a day via an automatic system that employed "electronic calculators." John Naisbitt, author of the 1982 bestseller *Megatrends*, predicted the "demise" of representative democracy in favor of a system where the people decide directly on "anything that comes up that impacts [their] lives."

Of course, technology could cut both ways when it came to the role of the citizen in the legislative process. In an increasingly complex world, where proposed laws could run to hundreds or thousands of pages, was it reasonable to expect the average person (who often is unable to name his or her two senators) to make an informed decision? Leland Hazard, an attorney and early proponent of public television, didn't think so. He believed that one day, unelected, highly trained technocrats would be running the complex machinery of modern civilization (although other civil liberties would be preserved). Robert C. W. Ettinger believed that there might be an electronic voting system, but before the computer allowed one to cast a ballot, the prospective voter would be tested on the issue. Fail the test, Ettinger believed, and you wouldn't be allowed to vote; do well enough, on the other hand, and the computer might allot you additional votes so your say counts for more. Technology, in short, was going to supersede the ancient principle of one man, one vote.

By the 1980s, the focus was on the more prosaic machinery of American government, a system that many believed was creaking and in need of change. John Naisbitt believed that the country was in the midst of a general trend where long-range thinking would be more vital than ever before; he advised job hunters of the time to look for companies that thought in terms of the next decade, not just next week. He also thought that this lengthening of perspective would spread to the political sphere as well—including an extended term for U.S. presidents of six years. Naisbitt and other forecasters saw more fundamental changes coming, too. He believed that America's two-party system of Republicans and Democrats was in its death throes, thanks to an electorate that was more and more likely to split the ticket and less likely to align itself with one party or the other (a group that today is referred to as the "swing vote").

A few were even predicting that America was going to get a little bigger. Back in the 1960s, when the American flag contained two freshly embroidered stars representing Alaska and Hawaii, it seemed as though the country had room for at least one or maybe two more. A 1968 forecast claimed that Puerto Rico, which had and has the rare status of being a self-governing territory of the United States, would make the leap to statehood. Through the years a number of votes have been taken in Puerto Rico—the last in 1998—asking residents if they support statehood. The answer has always been no. Statehood for Puerto Rico also encounters fierce resistance from some American conservatives, who point out that this fifty-first state would be the nation's poorest and is also primarily Spanish-speaking. This would not sit well in a country with an active movement to make English the official language.

Making major changes to the way America governs itself has proved to be a tough slog, partly because the politicians who would ultimately vote on such changes largely benefit from the status quo. Computer-aided direct democracy would upend Washington's lobbying system, for instance, if millions of people—not just the handful who currently represent them—had to be convinced to vote for that new highway in Congressman So-and-So's district. The closest we come to direct democracy today is the initiative and referendum system, where citizens, at least in some states (most notably California) can get certain questions on statewide ballots, such as whether to legalize medical marijuana or euthanasia for the terminally ill.

Chapter Nine

THE WORLD WILL END . . . PRETTY SOON

The Second Coming of Christ has been a long time coming. The original followers of Jesus believed that their messiah would return within just a few years after his death around 33 A.D. In the New Testament book of Mark, Jesus's contemporaries are told that "this generation will not pass away" before the Lord comes back and evil is vanquished. But even as the decades and then centuries came and went without a messianic return, Christians remained undeterred. Irenaeus, an influential bishop in Gaul (modern-day France) wrote in the late 2nd century that soon, Jesus would create his kingdom on earth and the dead would rise. In their entertaining book A *History of the End of the World*, Yuri Rubinsky and Ian Wiseman describe the mania that gripped Christian Europe in 1000 A.D., a millennium since the birth of Christ and hence a year seemingly freighted with cosmic significance. Thousands of European pilgrims made the arduous journey to Jerusalem, selling off their few belongings to await Christ's return right where it was supposed to happen (this, incidentally, also helped kick off the Crusades). Back then every calamity was seen as a sign that the end was nigh, from the invasions of the still-pagan Vikings to plagues and bouts of famine.

According to the book of Revelation—the part of the Bible that supposedly describes the end of the world and the dawn of a new epoch where Jesus would rule on earth—the messiah's return was preceded by the arrival of an "Antichrist," an arch enemy who would lead the forces of evil against the forces of good in a war that would demolish the old, sinful order. As a consequence, the history of the Christian world is peppered with references to the devil taking the form of various earthly figures. During the Protestant Reformation, the Catholic Church referred to Martin Luther as an Antichrist, and the Protestants returned the favor, labeling various popes Satan's agents on earth. Centuries later, in the early 1800s— an era when suspicion was otherwise giving way to empirical science—Napoleon was tarred with the Antichrist brush.

America, meanwhile, was becoming a hothouse for new Christian sects thanks to the Constitution's guarantee of religious freedom. Sometimes suffering from persecution or mockery, members of many of these sects were inclined to believe that, provided they adhered to their demanding beliefs, they would soon be saved and their enemies smote. In 1774, an Englishwoman named Ann Lee founded an offshoot of the Quakers known as the Shaking Quakers and headed for America. Followers were expected to remain celibate in preparation for the Second Coming, which was to happen in 1792. Another group, one that eventually became the Seventh-Day Adventists, predicted Christ's return with a remarkable degree of specificity—He was to come back October 22, 1844. The Jehovah's Witnesses may hold the record for the number of apocalyptic predictions made—about a dozen dates in the 20th century have been selected as the time when the end would begin. The present world was supposed to pass away in, among other years, 1918, 1932, 1975, and, of course, in 2000.

Underlying all these religions' predictions was, as was mentioned earlier, John's book of Revelation, which was written by a Jewish refugee living on the Turkish island of Patmos in the year 90 who described what he said was a vision imparted to him by God (hence the book's name). The book contains striking, complex, and varied imagery that might seem odd to a modern reader but would have resonated with John's target audience, 1st-century Christians in Asia Minor steeped in Old Testament tradition. The book opens with a description of a war in heaven that begins when a lamb, apparently a symbol of Jesus Christ, breaks the seven seals binding a scroll being held by God. With the breaking of each seal, a new calamity is unleashed, including the famous Four Horsemen of the Apocalypse (Strife, War, Famine, and Death). With the opening of the sixth seal, there is a terrible earthquake and the moon turns bloodred.

Later come the arrival of the beasts: the Whore of Babylon, a dragon with seven heads and ten horns, and a similar creature arising from the sea, who induces millions of people on the earth to follow him. (One or all of these creatures are commonly seen as symbolizing the devil.) This Antichrist briefly rules the world until he is defeated by an army led by Jesus, who then reigns over a paradise on earth for a thousand years.

Today most scholars believe that the Antichrist whom John was referring to was the contemporary Roman Empire, which around 70 A.D. waged a merciless campaign against one of its rebellious provinces, Judea (Israel). In a gesture of profound significance for the Jews and Christians (who at that point were still considered members of a Jewish sect), the imperial legions also destroyed the Temple in Jerusalem, an event supposedly predicted by Jesus in an earlier Gospel. Elaine Pagels, a professor of religion at Princeton University and author of *The Gnostic Gospels*, believes that the visions described in Revelation would have been a morale booster to the followers of the one true God, who at that time were being persecuted by the decadent, polytheistic Romans. Revelation gave Christians hope that those haughty, mighty emperors would soon— very soon—get their comeuppance, and that the Temple in Judea would be rebuilt. In one sense, then, the apocalypse has already occurred—the Roman "Antichrist" was defeated, although its collapse came about through strictly earthly means, not by the hand of God. The meek, martyred followers of Jesus prevailed in another sense as well—the Roman Empire had been Christianized for centuries by the time it fell apart around 455 A.D.

It wasn't until a few hundred years ago that academics began reading Revelation critically and in its historical context. But by that time the bulk of the Christian world was predisposed to see Revelation as describing events yet to come, no matter what a smattering of university professors had to say. And so it was well into the 20th century when a number of events occurred that, if anything, strengthened the belief that Christ should have come back to earth about a decade or so earlier.

The staying power of Revelation lies in both the richness and vagueness of its symbolism. Images like a dragon with seven heads and ten horns attempting to devour a pregnant woman who escapes after sprouting eagle's wings can be interpreted in more than one way, to say the least.

Still, 20th-century history did seem to jibe eerily with the cosmology laid out by John nineteen hundred years earlier. Most significantly, the founding of the state of Israel in 1948 was interpreted as one giant step toward the reconstruction of the Temple in Jerusalem, an event that to millions of Christians is to coincide with Jesus's return. (Even though Israel has been around for more than sixty years, though, the Temple isn't likely to be rebuilt soon—if ever. The site where it was located, the Dome of the Rock, is the second-holiest place in Islam.) Revelation describes wars where millions die in fire and earthquakes and terror rains down from the skies— a metaphor that could easily be applied to missiles carrying thermonuclear warheads. In the final half of the 20th century the Middle East was riven by strife, and more than one foreign policy expert believed that problems in the area could be the catalyst for an apocalyptic global conflict. The Middle East is home to the ancient battlefield Megiddo—from which is derived the word *Armageddon*.

Numerology also played into this interpretation. John speaks of a thousand-year reign on earth when Jesus does come back, so there's a certain pleasing, arithmetic symmetry in having that reign start precisely two thousand years after Christ first came to earth in the form of a baby born in a manger. The ten horns on the beasts were once believed to represent the nations that comprised the European Common Market (an interpretation that was complicated when that body grew to include more countries). Some also believed that the Antichrist would be born in June 1966, the sixth month of the century's sixty-sixth year, a date that could be represented by 666, "the devil's number" under the Judeo-Christian formulation. That would make the devil thirty-three years old in 1999, the year he would presumably have his final showdown with the forces of light and, in a satisfying coincidence, Jesus's age when he was crucified.

Skeptics and those outside the Christian tradition may scoff, but in the last decades of the 20th century there was an entire industry dedicated to the belief that the end times would begin on or close to the year 2000. Preachers in the 1970s, '80s, and '90s interpreted every other headline through the prism of millennial prophecy. The Persian Gulf War in 1990 was seen by some evangelists as the start of the Final Conflict, with Saddam Hussein in the role of the Antichrist. In 1998 the late Reverend Jerry Falwell produced a tape in which he suggested that the Y2K computer problem would spark a global breakdown that could "start a revival that spreads over the face of the earth before the Rapture of the Church." While stopping short of saying that Christ would come back in 2000, Falwell told his listeners "I wouldn't be surprised if He did."

A number of authors and moviemakers cashed in big on the millennial craze as well. The hit 1968 film *Rosemary's Baby* starred Mia Farrow as the unwitting mother of Satan. The book and movie *The Omen* had the devil's spawn reared by a prominent couple, again without their knowledge of his true heritage. Both these fictional demon babies would be coming into the fullness of their manhood at the start of the 21st century.

Perhaps no one had as much success disseminating pop Christian millennial belief than Hal Lindsey and C. C. Carlson, who wrote 1970's *The Late Great Planet Earth*, a publishing phenomenon that has sold nearly thirty million copies. Although Lindsey states in an introduction that he makes "no claim of knowing exactly when the world is going to end," just a few pages later he asks, "Will these predictions be fulfilled just as certainly and graphically as those of the first coming [of Jesus]? This writer says positively, 'Yes.'" The authors also note that these predictions would be fulfilled within a generation of the rebirth of Israel in 1948—in other words, sometime late in the 20th century.

The beginning of the end, the authors write, kicks off when various nations invade the young Jewish state, whose formation is blamed (or credited?) with ultimately "triggering the hostility which brings about a great judgment on all nations." The authors believed the Bible predicted that the Soviet Union would be a combatant in the millennial battle. They quote the Old Testament book of Ezekiel, which refers to a vast nation "in the uttermost parts of the north" (the Soviet Union was due north of Israel) fielding "a great host, a mighty army" that will make war against the Jewish homeland. Russia "is destined to plunge the world into its final great war which Christ will return to end," Lindsey and Carlson wrote. Given that many in the West were already concerned about "godless" Soviet power, this prediction must have packed quite a punch, being both frightening and plausible.

Russia was to lead a coalition that included sub-Saharan Africa and the Arab countries, as well as an Asian "horde" that would wipe out a third of the earth's population by "fire," "smoke," and "melted earth," as Revelation put it (a clear reference to nuclear war, the authors write). Revelations also described this eastern army as having 200 million soldiers. According to Lindsey and Carlson, the Chinese themselves boasted that they could field a militia of 200 million. "Coincidence?" the authors ask rhetorically—and ominously.

The authors even take a stab at identifying the Antichrist himself. Interpreting Revelation and other books of the Bible, Lindsey and Carlson argue that Jesus's nemesis would be the head of a European confederation, a charismatic leader who appealed to a generation that in the 1960s was "in to" occult subjects such as astrology and witchcraft. This great dictator would rule the world through economics, controlling a highly technological society via computers and barring anyone who resists him from buying, selling or holding a job (to this day, some fundamentalist Christians see Social Security numbers and bar-code, laser-scan technology as handy tools for a future Antichrist to keep his minions in line).

Ultimately, Lindsey and Carlson write, those who have kept the faith will be pulled up to Heaven in what is known as "the Rapture" prior to the great conflict involving the Russians, allied Europeans, and Asian "hordes." The authors go into great detail about the tactics that would be employed in this great final battle, including helpful tidbits like the fact that the Chinese would take an overland route as the Bible predicts because the nation would be too backward to possess enough ships and planes for all its soldiers. Their book even contains maps showing troop movements!

The success of Lindsey's books is all the more remarkable considering that a French druggist beat him to the punch by almost five hundred years. Michele De Notredame was also a bestselling purveyor of prophecy, making predictions about

the French nobility, the Catholic Church, and even technological developments, which won him considerable fame and
wealth. And that fame persists today, although he is now better
known by the Latinized version of his name, Nostradamus.

Nostradamus's predictions were written in the form of po-
ems called quatrains, some of which pertained to what was
to him the far future. He is credited by many as foreseeing,
among other things, the Kennedy assassination, the atomic
bomb, the 1969 landing of men on the moon, and the invention of the periscope. (Skeptics are quick to point out, though,
that the accuracy of Nostradamus's predictions is pretty much
in the eye of the beholder.) And, like Lindsey, he also foresaw
the start of the current millennium as the end of days. Unlike Lindsey, however, Nostradamus provided an almost exact
date:

"The Year 1999, seventh month, / A great king of terror
will descend from the skies, / To resuscitate the great king
Angolmois, / Around this time Mars will reign for a good
cause."

This quatrain was once interpreted by more than one
Nostradamus watcher as referring to end times. Stewart
Robb, author of the 1961 book *Prophecies on World Events*,
wasn't hedging his bet at all when he wrote that "we see Armageddon is on for 1999" based on this verse. "Terror from
the skies," Robb argues, is a clear reference to air warfare, a
remarkable prediction since it was made about four centuries
before the invention of the airplane. Mars was the ancient god
of war, and "for the good cause" means that the war is nearly
over, Robb believes. In another passage, Nostradamus writes
of "The Third Antichrist, soon annihilated / But his bloody
war will last twenty seven years." Since the war is just about

wrapped up by 1999, Robb wrote that this quatrain indicates that the great, final conflict had begun in the early 1970s. The first two of the three "Antichrists" are usually believed to be Hitler and Napoleon.

Robb was just one among a legion of Nostradamus interpreters who emerged in the decades leading up to the new millennium, and nearly all of those interpreters seemed to be loaded down with preconceived notions that greatly shaped the tenor of their opinions. So in 1980, a Frenchman, Jean-Charles de Fontbrune, concluded that a quatrain describing the "year the Rose flourished" referred to 1979, when the Socialists came to power in France (their party was symbolized by the rose). Nostradamus also predicted that this would be a year when Muslims and Christian powers would be in great conflict, and, in a sense, it was—1979 was also the year that the Iranians stormed the U.S. embassy in Tehran and took several American hostages. Even more alarming, de Fontbrune said his reading of Nostradamus also indicated that late in the 20th century, Islam would destroy the Catholic Church, Arabs would team up with the Soviets to invade Western Europe, and Paris would "swim in blood." De Fontbrune's writings reportedly set off a mini-panic in France that year.

Millennial prophecy was hardly confined to the extremes; mainstream clerics such as Billy Graham also fueled the year 2000 hype. Graham, the spiritual adviser to a long succession of U.S. presidents and at one point the world's most famous Christian evangelist, counted no fewer than twenty-two signs that the end times were nearly upon us, he told an interviewer in the early 1980s. Included among these signs were a supposedly rising tide of devil worship (evidenced by the popularity of occult-themed books such as *The Exorcist*) and an upturn in the number of UFO sightings. The Catholic Church is also reportedly sitting on its own end-of-days prophecy, one supposedly conveyed to a group of small girls in Portugal in 1917 by a vision of the Virgin Mary. Although the church has refused to officially confirm anything, reportedly the vision told the girls that the world was due for a fiery end during the term of the fifth pope after the prophecy was revealed in 1960, Rubinsky and Wiseman write in *A History of the End of the World*. As of this writing, only the fourth pope, Benedict, now holds the office—so it's too soon, technically, to debunk this particular prediction.

The forecasts about a 21st-century end of days weren't just limited to people guided by the Bible, however. Psychics, astrologers, and assorted others who dabbled in the New Age arts also believed that we were heading for a turning point at the change of the millennium. Among the more famous of these was Richard Noone, who in 1982 published a book called *5/5/2000 Ice: The Ultimate Disaster*, which pretty much delivered on its title, describing an earth-changing catastrophe that begins on the day, May 5, when several planets in the solar system would be in alignment. According to Noone, this unusual but hardly unprecedented event—human civilization had somehow survived a similar alignment in 1962—would trigger a massive solar flare, disrupting earth's magnetic field and dislodging trillions of tons of ice at the South Pole. The result would be massive flooding, earthquakes, volcanic eruptions, and all the other ingredients of a really good, scary apocalypse. In making his case, Noone strays just a little from conventional scientific argument—for instance, he claims that a civilization older than the ancient Egyptians really built the pyramids, and they did so explicitly to warn humankind of the May 5, 2000, disaster.

Gordon-Michael Scallion was another forecaster who believed that the earth was due for a shakeup in the late '90s or early '00s. A self-described futurist who claims to have traveled back in time to see how the pyramids were built, Scallion has made a career out of predicting earthquakes, hurricanes, and other natural disasters that have a habit of not occurring when and where Scallion has said they will. He also sells "future maps" showing what the world is going to look like after these disasters. In one, California is just a string of islands and Denver is on the coast of the Pacific following an earthquake that was to occur in 1998.

Scallion believed the few years around the millennium were going to be especially active, and at one point predicted that the years 1998 through 2001 were going to see the most dynamic changes in the earth in eleven thousand years. Massive solar flares were going to paste the planet, causing massive disruptions to the power grid. Even more incredibly, new species of animals would emerge thanks to genetic mutation. In 1999, the world's governments were to confirm the existence of UFOs. And no self-respecting psychic could possibly discuss the millennium without making some kind Revelation-inflected call. Scallion claimed the final battle between good and evil was going to take place in Turkey.

Another psychic named Terry Tim Rodgers predicted in 1999 that Pope John Paul II was going to die that year (he actually passed away in 2005) and that the Antichrist (him again!) would come into power shortly before January 1, 2000. Rodgers also helpfully gives us the evil one's birth date and ethnicity—he is a Middle Easterner who was born November 12, 1967. Several others predicted that the mythical continent of Atlantis would reemerge from the depths in the late 1990s or early 2000s, a forecast first popularized by Edgar Cayce, the man regarded as the father of the American New Age movement. A survivalist group in Arizona believed that in 2000 the pyramid at Giza would be opened, revealing the "secrets of the world."

Sadly, the millennial mania also claimed some real victims. In March 1997, authorities discovered the bodies of thirty-nine people at a California home who belonged to the Heaven's Gate cult, a group that mixed elements of religion and science fiction in their beliefs. The group's leader, Marshall Applewhite, convinced his followers that the earth was about to be "recycled," the arrival of the Hale-Bopp comet signaled the return of an alien spaceship, and that those who renounced their earthly bodies would survive the apocalypse to join these interplanetary visitors. Among Applewhite's followers were people who left behind families and good jobs, and even some men who reportedly castrated themselves to demonstrate their devotion to the cult's beliefs.

The simple fact that a string of end-of-the-world dates during the last few thousand years have come and gone—with the world still intact—has not deterred those who are convinced an apocalypse is imminent. Many Christians, especially in America, still believe the end times are coming soon—although most no longer try to suggest a specific date.

But yet another crop of end-timers have emerged in recent years, and this group believes that the world as it now exists will last only until late December 2012. Around this time a number of things will happen all at once: the ancient Mayan astronomical calendar, which was discontinued during the Spanish conquest of Mesoamerica, comes to the end of a more than 5,100-year cycle. Also around this date, the earth will come into alignment with the center of the galaxy.

Thanks to these two events, it was perhaps almost inevitable that this time period would come to be regarded as yet another "apocalyptic" date (never mind that there's no evidence to suggest that the ancient Mayans themselves considered this date anything more than the end of a calendar cycle, akin to the start of a new century). At least four bestsellers have been written in the last few years pegging this date as a time when the world will end, possibly due to environmental collapse, or, more hopefully, as a time when a balance will finally be struck between "the masculine and the feminine," as Andrew Smith forecasts in his book, *The Revolution of 2012*. Spend a few minutes with a computer search engine and you'll discover that some forecasters have linked this latest millennial bash with old standbys such as Atlantis, the Egyptian pyramids, and Nostradamus. At the risk of tempting fate, I personally think that this family of predictions will also find its place on the growing ash heap of incorrect end-times predictions. However, if an asteroid does start to speed toward earth in December 2012 (as some of the more pessimistic forecasters believe), then I'll be the first to apologize—provided, of course, I have the time to do so.

It's human nature to worry about the future, which, if this book has shown anything else, is largely unknowable. Believing in an apocalypse is a way of imposing a little certainty on the world of tomorrow, a way to get oneself psychologically prepared for what's to come. (Even if what's to come is pretty terrible, knowing about it at least gives us a chance to brace ourselves.) And although the early 21st century was supposed to see the end of time in the Christian world, the concept of

Armageddon also appeals across cultures and the ages. As a species we are almost congenitally disposed to believe that one day, everything will end.

It's a valid belief. Science has shown that in the last few billion years there have been a number of global catastrophes, such as asteroid impacts and abrupt shifts in climate, that have wiped out virtually all life on this planet. And one day, billions of years from now, the sun will ultimately die and the earth will become a barren, deserted rock. There's a reason why we have such a deeply ingrained fear that, maybe, this time, the end really is nigh.

But we humans are natural optimists, too. After all, one of our most powerful instincts is to procreate, to make children who we believe will see tomorrows that we never will. Most parents probably hold in their hearts some hope for the future, or else they likely wouldn't bother helping to build that future through their offspring.

This duality in our nature reflects something that's true about life: sometimes tomorrow is better, sometimes worse, sometimes about the same. It's this very uncertainty that makes life worth living.

AFTERWORD

Right on the Money

Throughout this book, we've seen forecasts that have gone awry for a variety of reasons. Sometimes, prognosticators assumed, wrongly, that just because something is a good idea means that it will get done. Doesn't mean it's going to get done. For a lot of reasons, for instance, making cars that no longer run on gasoline has been doable for quite some time; there are powerful economic and environmental reasons for us to disconnect the petroleum IV tube that our society is hooked to, and those reasons have been known for decades. Yet only now, almost forty years after the first oil shocks, are we even coming close to building cars that run on something other than the ancient, earthly remains of dinosaurs. When trying to figure out what's coming next, common sense isn't always the most reliable guide.

Another fallacy that complicates the business of prediction is what I'll call the incremental error. This is the assumption that simply because we're close to a breakthrough, we'll soon see that breakthrough. Sometimes the longest part of a journey is at the beginning, but sometimes it's at the end. Take nuclear fusion, for instance—scientists have been at the doorstep of producing electricity from this source for forty or fifty years, and for that long, serious people have believed we were about to cross the threshold. But so far, fusion from power remains one of those advances that is perpetually "just around the corner." The same could be said, for slightly different reasons, about the flying car.

My point here is that though I've had fun describing some of the loopier forecasts from years past, I still do have a lot of respect for the people who made the attempt. And in that vein, I think it's time to give credit where credit is due.

For all those predictions about 21st-century flying cars, genetically engineered designer babies, and three-day workweeks, there were also a number of guesses that were stunningly prescient. Many of the futurists we've met in this book also demonstrated an uncanny knack for discerning the shape of the future by correctly, sometimes counterintuitively, assessing emerging trends decades before they came to fruition.

And a few forecasters even saw hundreds of years ahead, accurately predicting developments by relying solely on the power of their insight. John Wilkins, an English clergyman with a strong scientific bent, predicted in his 1640 book, *A Discourse Concerning a New World and Another Planet*, that human beings would one day journey to the lunar surface using "flying chariots." "'Tis likely enough that there may be a means invented of journeying to the Moon; and how happy they shall be that are first successful in this attempt." What's especially remarkable about this prediction is that when it was made, accepted wisdom was completely wrong about the nature of the solar system. Just a few years earlier, Galileo was placed under house arrest for daring to suggest that the planets revolved around the sun and not vice versa. He had already been reprimanded for challenging the belief that the moon itself was not a perfect sphere but was pocked with craters.

Another medieval prognosticator, Roger Bacon, the 13th-century English philosopher, imagined a time when human beings would build flying machines moved by an "unspeakable force" "without any living creatures to stir them" (Bacon was writing when the prime source of locomotion was the horse and the ox). This device would have an "engine, by which the wings, being artificially composed, may beat the air after the manner of a flying bird." Bacon made this prediction about five hundred years before human beings first flew in hot-air balloons and about seven hundred years before the invention of the airplane.

A few centuries later, during the 1790s, a Frenchman, the Marquis de Condorcet, produced what may be the most stunningly accurate string of predictions ever recorded—stunningly accurate because much of what he described was not to take place for at least a century.

Condorcet, who lived through the French Revolution, believed that monarchies and established orders across the globe would be toppled by popular revolutions (how many 20th-century governments have been destroyed when the "peasants" forcibly threw their leaders out?). In a similar vein, he believed that eventually, European nations would lose all their colonies in Africa, Asia, and the New World (a process that wasn't complete until 1999, when the Portuguese relinquished control of Macau). When Condorcet was writing, European colonialism was still on the upswing—it would be several decades before the British Empire, to take the best example, would reach its maximum extent.

Writing at a time when most people were self-sufficient farmers who received little from the state except demands for more taxes, Condorcet also foresaw the creation of a government system guaranteeing financial support for orphans, the elderly, and widows—something very much like today's Social Security. He predicted the rise of universal schooling, which would propel yet another trend—a dramatic increase in scientific knowledge and technological skill. Thanks to these advances, "A very small amount of ground will be able to produce a great quantity of supplies of greater utility or higher quality" than was presently possible, he wrote. And today, of course, farms are colossally more productive than they were in the 18th century.

At a time when outbreaks of cholera, malaria, and other infectious diseases were routine, Condorcet accurately predicted that medical science would lick most of these illnesses and "the average length of human life will be increased . . . [and] better health and a stronger constitution will be ensured." The average life expectancy in France at the time was about thirty years, much less than half what it is today.

This was also a time when women had virtually no rights, where a wife was, under the law, virtually the possession of her husband. Condorcet saw what was to come more than a hundred years hence—legal equality between the sexes, including ensuring women the right to be educated.

Another Frenchman is also remembered as an especially adept futurist: Jules Verne, the 19th-century author who is widely credited with creating the science fiction literary genre. And to this day he remains one of the best, thanks largely to his almost spooky ability to foresee in breathtakingly accurate detail some major 20th-century technological innovations and natural discoveries.

In 1863, Verne wrote a book titled *Paris in the 20th Century*, describing a world that was so fantastic that his friend and publisher, Pierre-Jules Hetzel, advised him to wait twenty years before releasing it because "no one would believe" what Verne was describing (the manuscript was forgotten until it was discovered in a safe in 1989). Verne, who was somewhat pessimistic about the effect technology would have on mankind, wrote about a man who comes to a tragic end in a Paris that contains steel and glass skyscrapers (twenty years before the first modern one was built in Chicago), air-conditioning (it would be forty years until electric air-conditioning appeared), gas-powered cars, and electronic calculators. In this world there is also a global communications network, long before people could make international phone calls. In other works Verne predicts the arrival of television news in color, the helicopter, and trains that operate under the Atlantic (the "Chunnel," the train running beneath the English Channel and linking Britain and France, comes pretty close).

One of his more famous works, *From the Earth to the Moon*, contains several details about the first lunar journey almost a century before it occurred. Verne describes a voyage being undertaken by three travelers (there were three men on the *Apollo 11* mission) who are sent on their trip from a Florida town about one hundred miles from what would later be known as Cape Kennedy. Their journey concludes with a splashdown in the Pacific. He describes the sensation of weightlessness that his travelers experience after they leave earth orbit. The novel also describes how animals were sent up first, before the humans, to ensure that space travel was safe for living creatures. In the late 1950s and early '60s, chimps were indeed America's first space travelers; a Russian dog, Laika, was sent into orbit four years before the first cosmonaut did.

Verne's near contemporary and another giant of science fiction, H. G. Wells, was considerably good at the futurism game, too, making accurate predictions not only about earth-shattering events but also about the quotidian details of modern-day life. He envisioned a time when "whole regions will be given over to opulent enjoyment"—a description that roughly applies to such contemporary vacation destinations as Las Vegas or the Riviera. He also envisioned something like today's celebrity culture, when the kinks and peccadilloes of the rich and famous "will be known of, thought about and more or less thoroughly discussed by an enormous and increasing number of the common people." Coinciding with this would be a general easing on moral restrictions, Wells believed, "when freedom of escape from disapproving neighbors will be greatly facilitated."

No facet of life was too small for Wells's forward-looking gaze. Writing at a time when guys still wore ties and wool jackets even when they went on picnics, Wells predicted that by now, "men will appear in the elaborate uniforms of games" (consider all the adults out there today whose wardrobe is festooned with team jerseys and baseball caps). Women, meanwhile, were going to go for the retro look: they will "ransack the ages for becoming and alluring anachronisms," he wrote. He predicted that household conveniences would eradicate the "servant class" (at this point in the West, the homes of the middle class and the wealthy normally included rooms for live-in maids, butlers, and cooks). He believed that businesses would leave the city center and move to the suburbs, which would be linked by broad highways teeming with cars.

He was also among the first to foresee the rise of nursery schools for young children. He believed that the 20th century would see general relaxations against sexual taboos, leading to a rise in out-of-wedlock births. As women asserted themselves more, divorce would spike as well.

And he accurately anticipated many of the horrors of 20th-century warfare. In *Anticipations*, he wrote that future wars would be won not by the side with the bravest soldiers but by the nation that had the most advanced weaponry and best scientists. In 1908 he envisioned a time when cities would be annihilated by aerial bombardment. Before World War I, he foresaw the time when governments would take over vast sections of their economies and marshal them toward war production; the energies of an entire society would be needed to make future war.

Perhaps his most chilling prediction came in his 1914 novel, *The World Set Free*, in which he described "atomic bombs"—Wells may very well have been the first to use the term—fueled by a man-made element and capable of destroying entire cities. The bombs were dropped by airplane and created "volcanoes," a vast conflagration of fire that is a pretty good approximation of what an actual nuclear detonation is like. Not for nothing is Wells considered by many to be the father of future studies.

A few decades after Wells's death in 1946, the art of predicting the future was struggling to become a science, a methodology that employed statistical analysis, game theory, and computer-aided forecasting, not just savvy guesses. Two of these more scientific forecasters were Kahn and Wiener, the authors of *The Year 2000* and whose erroneous predictions are found throughout this book. Their methods, though, did produce some astute guesses as well.

At a time when the possibilities of the recently invented laser were just barely understood, the two foresaw a big future for this device, not just as a weapon of war but as a tool of commerce (if you shop anywhere with an automatic price scanner, drive through an automated toll booth, or have corrective eye surgery, you are benefiting from laser technology). The two believed that modern medicine would provide drugs that would control a range of mood disorders, help boost a person's ability to concentrate (drugs that are prescribed for hyperactivity and attention-deficit disorder achieve this), as well as easier-to-use birth control. They also predicted the rise of the ATM culture, with computerized, twenty-four-hour banking and credit systems—not to mention the rise of pagers, "pocket phones," and inexpensive home video recording. They even predicted something like today's online universities, stating that "computerized and programmed learning" in the home would be commonplace by the 21st century.

Although more a journalist than a scientific forecaster, Alvin Toffler also had some ability to see around the corners; his observations about emerging societal trends are especially good. When his book *Future Shock* came out about forty years ago, for instance, many people tended to be joiners, taking part in civic activities and organizations such as the volunteer fire department or the Elks. Today, however, people are more likely to nest in their own homes than join (a modern phenomenon explored in books like Robert Putnam's *Bowling Alone*). Back in 1970, Toffler noted that more and more people, feeling less of a sense of being rooted to place, were becoming less likely to participate in local elections or attend town meetings. He observed a similar emerging phenomenon in the workplace, where, Toffler suggested, the old idea of spending one's entire career at the same company was giving way to one where workers were hired only for specific projects and had diminishing loyalty to the firm. Long the case in America, even in today's Japan—where the "salaryman" once spent decades working for the "team"—young workers expect to change jobs several times during their careers.

Toffler also accurately foresaw the rise of new kinds of family at a time when there were powerful taboos against certain kinds of relationships. He believed that one day it would be socially acceptable for divorced fathers to win custody of the kids, a phenomenon that's still not quite the norm today but is a lot more common than it was back in the '70s. He also predicted that there would be more "aggregate" families, consisting of remarried divorcees and the children from their first marriages. Even more incredibly, Toffler believed that gay couples would one day be permitted to adopt children. This was at a time when homosexuality was still considered a mental disorder in America, where the police could "roust" gay nightspots with impunity, and where the election of the nation's first openly gay public official was still a decade away.

Another futurist we met earlier in this book, Ferdinand Lumberg, also had a succession of smart guesses, particularly regarding the U.S. economy. Unlike many others at the time, Lumberg said "it was hardly likely" that the workweek would be substantially reduced from the forty-hour standard, although he did accurately predict that automation and other factors would lead to a decline in the number of blue-collar workers—exactly what has occurred in modern, postindustrial economies such as America and Western Europe. In a related development, Lumberg accurately foresaw labor unions declining in size and power as well. In the early 1960s, about a quarter of America's labor force was unionized. Today less than 9 percent of private-sector workers belong to a union— the lowest level in seventy-five years.

With well-paying blue-collar and clerical jobs scarce, many more people, Lumberg predicted, would enter careers "requiring elaborate education"—and indeed, today there is a direct correlation between the amount of schooling someone has and his or her income level. Workers without special training, meanwhile, were to "sink into the mass of the dependent unskilled," he wrote. The average wage of someone with a high school education or less has sunk like a stone since the 1970s, precisely what you would expect in the world Lumberg described forty-five years ago.

Lumberg also accurately predicted a vastly changed media landscape, with the daily newspaper industry experiencing a steep, long decline—although here Lumberg was hardly alone. Some prognosticators correctly foresaw mass media becoming a lot more fragmented than it was in the days where a major city's news outlets consisted of just one or two newspapers and three television channels. Long before the advent of modern communications equipment, back in the 1950s, the physicist A. M. Low predicted that there would be a multiplicity of radio stations, many serving niche markets—exactly what has happened, where satellite radio has hundreds of channels, some of which are dedicated to just one artist. Writing a decade later, Theodore Gordon anticipated something similar happening with television, which would get a hundred channels. (He even predicted that certain channels would be dedicated solely to covering the financial markets and that buying stock would become an automated process.)

In 1968, Ithiel de Sola Pool, chairman of the political science department at the Massachusetts Institute of Technology, surveyed the future media landscape and was able to make some remarkably perceptive calls. When Pool was writing, TV viewers had to take what was dispensed to them, when it was dispensed to them. But Pool believed that today, video recording equipment and multiple television channels would enable us to pick and choose which narrow form of programming we desire, which we watch "at our own convenience," with computers helping us make our choices. Modern recording systems such as TiVo do in fact use computer technology to help us keep track of our favorite shows and to watch them whenever we like. He even described something like YouTube, the online video service: in the future, "every little clique of creative people makes its own movies incorporating its own idiosyncratic point of view."

Others accurately foresaw that this was going to wreak havoc with the traditional broadcast networks, NBC, CBS, and ABC, which have been forced to buy up a few niche cable channels in order to stay competitive. "National network television is in for a long, downward slide. . . . My own guess is that by the end of the decade, [the networks] will have half the viewers they do today," John Naisbitt wrote in his 1984 book, *Megatrends*. The major networks have indeed lost a significant share of their old viewership, a trend that even today shows no sign of reversing.

Another futurist we've encountered throughout this book, Fritz Baade, also had his fair share of quality forecasts. In the early 1960s, when many experts assumed that rising prosperity would lead to massive population growth, Baade accurately foresaw what people across the globe actually would tend to do when they had more money—have fewer children, but invest more in the ones they do have. "[The] assumption that poverty checks population growth simply does not square with the facts—which, on the contrary, prove that population growth is checked far more effectively by higher standards of living. . . . [Families'] inclination and capacity for intelligent family planning tend to increase in proportion to their economic welfare." Today the wealthiest countries in the world are in fact those with the smallest average families, while in developing countries it's not uncommon for people to have five or six kids, even if they can't adequately provide for those children. As the world has, on balance, become wealthier, population growth has indeed been slowing, according to U.N. estimates.

And contrary to the many doomsayers who believed that the 21st century would see a massive food crisis thanks to huge population growth, Baade argued nearly fifty years ago that the world's farmers were up to the task. In his book *The Race to the year 2000*, Baade notes that around 1800 the British political economist Thomas Robert Malthus also predicted that the number of people would soon outstrip the food supply—an assumption that was as wrong then as it is today. Between 1800 and 1900, Baade notes, the population of England and Wales grew from 9 million to 34 million, yet the average person in 1900 consumed twice as much meat and four times as much sugar as his ancestors did. Baade believed, correctly, that even more food would be available in the 21st century.

The people who guessed correctly about future developments certainly deserve a lot of credit, but so do the people who attempted to foresee fifty or a hundred years ago what today would be like—and got it absolutely wrong. In helping us understand the limits of forecasting future events and trends, these men and women helped refine the art of futurism. As Thomas Edison once said about his repeated attempts to perfect a new invention, "I haven't failed, I've just found ten thousand things that don't work." In that sense, none of the prognosticators we've met in this book could be described as having failed.

ACKNOWLEDGMENTS

This book was several years in the making. Fortunately, though, I had the help and encouragement of a number of talented, dedicated people who took time out of their busy lives to assist in the production of this manuscript.

Cindy Capitani, Mayre Milo, and Alicia Zadrozny all read and reviewed portions of this book early on. So too did Darren Cooper, Jeff Jerista, and Steven DeVries, all of whom offered particularly insightful advice. Anje Kastl and Jeff and Stacey Clark provided invaluable technical assistance.

I would also like to thank the staff at the main branch of the New York Public Library and at the library of Montclair State University, two institutions where much of the research for this book was conducted.

I have frequently been asked for advice by aspiring writers on how to transform their idea for a book into a paper-and-ink reality. Consistently, that advice has been to check out what Mediabistro has to offer. With a full menu of online services, as well as in-person offerings in some cities, Mediabistro really is a one-stop shop for anyone looking for guidance in the world of book publishing.

It was through a class at Mediabistro's main offices in New York City that I met my agent, Ryan Fischer-Harbage, who I would also like to recognize here. Ryan worked with me, taking a rough idea and honing it into a proper manuscript, as well as helping me—a first-time book author—launch what I hope will be a long and fruitful career. This book would not have been possible without his dedicated efforts.

Finally, I'd like to express my gratitude to my editors at HarperCollins, Melissa Bobotek and Matt Harper, who both helped make this book much better than it would have been otherwise.